INSIDE THE THIRD WORLD VILLAGE

This book is about everyday life in a contemporary Egyptian village and the struggle to survive under extremely hard economic conditions.

Over the last two decades many of the men in the village have spent long periods working in the Gulf. The villagers see this as the most viable option to make ends meet. The issue of labour migration is seen to be a complex one, impacting as it does on a variety of levels – a globally organized labour market, the village, peasant households, as well as individuals.

For those left behind, generally women, the young and the old, subsistence agriculture continues. Wider responsibilities in the household, the money their husbands earn abroad and new images of the perfect peasant women are only some of the resources that women make use of to realize their own aspirations and ameliorate their living conditions.

Inside the Third World Village presents a detailed exploration of the changing assumptions and expectations of these villagers, of the ways in which they have grafted the demands and opportunities of the developed world onto the community of the underdeveloped village.

Petra Weyland is a researcher at the Orient Institute of the German Oriental Society based in Istanbul.

INSIDE THE THIRD WORLD VILLAGE

Petra Weyland

ROUTLEDGE

London and New York

First published 1993
by Routledge
11 New Fetter Lane, London EC4P 4EE

Simultaneously published in the USA and Canada
by Routledge
29 West 35th Street, New York, NY 10001

Typeset in Garamond by J&L Composition Ltd, Filey, North Yorkshire

Printed and bound in Great Britain by
Biddles Ltd, Guildford and King's Lynn

British Library Cataloguing in Publication Data

A catalogue record for this book is available from the British Library

Library of Congress Cataloging in Publication Data
Weyland, Petra, 1954–
Inside the Third World Village / Petra Weyland.
p. cm.
Includes bibliographical references and index.
ISBN 0–415–08851–8 (hard) : $55.00
ISBN 0–415–08852–6 (pbk.) : $16.95
1. Egypt—Rural conditions. 2. Alien labor, Egyptian—Persian Gulf Region. I. Title.
HN786.A8W49 1994
307.76'2'0962—dc20 93–1173

ISBN 0–415–088518 (hbk)
ISBN 0–415–088526 (pbk)

CONTENTS

ILLUSTRATIONS

PLATES

MAPS

TABLES

ACKNOWLEDGEMENTS

'We are *ʿishariyyin!*'

Im Rida, the aged mother of Rida, is sitting at the front door of Sahar, the vegetable pedlar. Joking and shaking her body in rhythm she sings a few lines from a song by Laila Nazmi:

> *ʿIshariyyin*, oh friends, we are *ʿishariyyin*, we ask after the one who asks after us! Never will something shameful come from us, we understand each other and we take our time. . . .

ʿIshari signifies 'to be intimate with', 'to associate with', it is an expression for sociability and for conviviality, for friendliness. This is in fact how I came to know the villagers. My choosing of Kafr al-ʿIshra, the 'Ishra village', as a pseudonym for the village is not only to guarantee anonymity but also to cordially express my gratitude to the village's women, men and children for including me, the stranger, in their *ʿishra.* I owe my greatest thanks to the ʿOmda's household, to Janette and Nafisa.

Elisabeth Steiner, Börte Sagaster and Angela Zerbes emerged in the final stage, when I needed help most – to them I express my appreciation. I also want to thank Christoph Reichert and Gabriele Sturm who patiently helped me with statistical operations. Finally, my thanks go to Sharon Siddique and Helmut Weber whose initial encouragement and support were vital for getting this study off the ground.

1

INTRODUCTION

Without any doubt, migration is an empirical fact of tremendous importance in the contemporary world. Hardly a day passes without the media reporting about illegal immigrants; about refugees for economic, war, civil war, poverty or even environmental reasons; about asylum-seekers, migrant workers, the brain drain. Reports abound about the ever-growing number of women and men who travel over long distances in a daily, weekly or even longer rhythm, because no suitable job or no work at all is available near their living place. We are aware of increasing numbers of persons from North Africa trying to enter Spain via the Straits of Gibraltar, and that many of them die on overcrowded boats; we learn that Somalian 'boat people' are risking the same fate off the coast of Yemen; we realize that increasingly draconian measures are implemented to ward off illegal immigrants at the US southern border; in Central Europe we now and then can follow the drama of second and third generation immigrant children who do not want to return home to their parent's native communities in Turkey or North Africa; we suddenly notice that we have already forgotten the large numbers of ethnic Germans from East European countries claiming a right to settle in Germany as media and politics have changed their focus to lamenting the floods of illegal asylum-seekers pouring into the country; we equally realize that the agonizing pictures of Albanian refugees in Italian port cities, and those of Arabs and Asians displaced in the Jordanian desert by the Gulf War, have long since been replaced by even more horrible photos of citizens of what had formerly been Yugoslavia. Media images and messages are of this type and colour the global public's reception of international migration.

However, there are also other migrant flows which, if reported at all, are usually much less spectacularly presented to the international public. There are the men from the black homelands working in South Africa's mines, women from the Philippines employed as housemaids everywhere in the Middle East, Indonesian workers seeking access to the labour market of prospering Malaysia; and there are also the western experts working as consultants and managers in developing countries. At the close of the twentieth century the list of people on the move is endless and getting longer every day. The United Nations Development Programme (UNDP) indicates that there were 75 million migrants worldwide at the beginning of the 1990s; other estimates were as high as 80 to 100 million even in the early 1980s (Khan 1989). No region of the world today seems to be excluded from migratory movements. If the world has become a single place, then migratory global flows are definitely among its most impressive expressions and one of its constituent elements at the same time.

Subsuming all these empirical forms of migration – to which others of the past would have to be added – under one definition or model seems too difficult, if not impossible. Spatial movements vary considerably in distance, at times covering only some hundred kilometres, at others thousands, crossing or not crossing international frontiers, leading from rural to urban areas and from towns to cities. And, as I already have stated, reasons for migration are also numerous. The landscape of contemporary global migratory flows gets even more confusing when we watch these flows constantly shifting direction and changing in magnitude and quality. According to the most widely circulated approach societies involved in migration are characterized as either 'migrant sending' or as 'migrant receiving', and there are also countries qualifying simultaneously as a so-called 'push-' and 'pull-pole'. In this latter case out-migrating workers are replaced by in-migrating (cheaper) labourers. While refugee migrant flows in particular are often directed to regions which are very weak economically (as, for example, Sudan or Yemen), a considerable part of the migratory flows is directed towards the world's centres of capital, production and services: migrants, even if not classified as labour migrants but as refugees or asylum-seekers, need jobs and it is at such places that one usually has the best chance to find one.

Efforts to come to terms with this bewildering diversity of contemporary and past forms of migration can take two directions. One would be to construct a highly abstract model, a kind of paradigm into which all these empirical manifestations could be fitted – elements of such a paradigm have just been alluded to in the foregoing paragraph. This paradigm is usually substantiated by introducing statistical data, often derived from official institutions such as border authorities, and from national census data – conventional statistical data about the magnitude of migration and the directions it takes, about the spending of remittances and about the main demographical and numerical characteristics of migrants – which are compiled and processed. Dealing with such figures and macrostructures is, of course, important as in so doing an overview of the numerical importance of an empirical fact is provided. However, my problem with such procedures is that explanations of that sort ('migration is a process whereby a person is pushed for some reason from one pole and for some other reason attracted to another pole'), in their claim for overall validity, are so general and abstract that they therefore risk the fall into banality. One other major objection is that such highly abstract paradigms tend to lose sight of their very objects: the migrants, who, after all, are human beings.

Hence, another path to clarification consists in scrutinizing, as closely as possible, one specific empirical form of migration and producing a case-study. It is only by engaging in such a project that the migrants themselves, their families and the communities they are part of, emerge from behind abstract concepts and structures derived from highly sophisticated statistical procedures thus enabling us to reach an understanding of the relevance of migration to their practical everyday lives. This, to my mind, is one indispensable aspect if we want to get an idea of the overall functioning of the global migratory system. To me, it is essential to understand labour migration in an encompassing, dynamic and historical framework which at the same time does not reduce the migrants to the sole capacity as carriers of labour power but rather sees them for what they are – as human beings, integrated into social networks and involved in making sense of the world in which they live. The analysis of macro-structural data is important as thereby a framework is provided for the understanding of a specific empirical setting on the micro-level, such as the village community, the peasant household, the individual person.

Having made these preliminary considerations I now want to introduce one such empirical case. Delineating labour migration to the oil-producing Gulf states I shall eventually focus on the migration of Egyptian peasants to these countries.

Since the early 1970s the oil-producing Gulf states and Libya have attracted particularly high numbers of foreign, especially Arab and Asian, labourers. In the wake of the oil crisis of 1973 oil prices had gone up tremendously and allowed producing societies, especially Saudi Arabia, Kuwait, Abu Dhabi, Iraq, the UAE, but also Libya, to engage in ambitious infrastructure projects which could only be carried out by the massive import of a foreign labour force. Nearly all the Arab countries have been involved in these migratory flows instigated by the oil business – either as migrant-sending or as migrant-receiving, but also as simultaneously receiving and sending societies. Major sending economies have been those of Syria, Lebanon, the West Bank and Gaza, Egypt, and Yemen. Iraq, Jordan and Oman have witnessed both in-migration as well as out-migration. Examples of the rapid change in numbers and directions of global migration flows are Kuwait and Iraq which as a consequence of the Gulf War sent back virtually all of their foreign working staffs. Yemen is another case in point, as nearly 1 million workers returned home in the wake of the Gulf crisis; it might, however, turn into a labour-force-importing country soon, as substantial oil resources have recently been found and are now beginning to be exploited. But in another respect the situation in Yemen is telling: poor and labour exporting as it has been until now it had nevertheless been importing considerable numbers of Chinese workers who were employed in road construction. The impression, therefore, that migration across national boundaries in that region of the world is not restricted to Arabs is correct: Asians form an increasingly large percentage of the expatriate work-force in the Gulf. But contrary to the Arabs, who usually migrate individually, these Indians, Koreans or Pakistanis are employed in groups, and although the oil extracting industry has attracted foreign workers, most of them have found jobs in other branches – particularly in the construction sector.

In the first decade of the oil boom especially, huge infrastructure programmes were implemented, and highways, airports, schools, universities, hospitals, and living space were built by foreign hands. Expatriates also crowded the service sector, working in shops, gas stations, restaurants, or as drivers. But not all migrants are unskilled and work in lowly paid jobs. University graduates from Palestine, Egypt or Syria are to be found in the higher echelons of the service sector, working as teachers, university professors, medical doctors and even in government bureaucracies. At the opposite end of this hierarchy of foreign workers we find those who are dependent on the daily wage labour market. Here again, the construction sector is most important. Women, both Arab and Asian, also work in these countries, mostly as teachers, medical staff, house maids or nannies – the latter sometimes outside the Gulf countries. I have seen rich Gulf Arab families in the waiting lounge of an international airport, their women covered in black and heavily decorated with gold, sniffing at the little bottle of 'Poison' just purchased in the duty-free shop; the men busy with the whole group's tickets, passports and boarding cards; children dressed in jeans and the fashions of the west; and a young and silent Philippine woman in their retinue occupied with the family's toddlers and youngest children.

Briefly stated, the foreign labour force working in the Gulf is relatively heterogeneous. We find expatriates occupied in a large range of qualified and unqualified jobs in the agricultural, industrial and the service sectors, and we may notice that many of them do not even stay a year while others are on their way to establishing stable migrant communities as their wives and children have followed them. Here, the Palestinians, already emigrants since before 1948, definitely form the most salient example of relatively prospering and stable migrant communities.

It would certainly be useful at this point to come up with some statistics[1] to pinpoint the magnitude of this oil migration. Yet I am very hesitant to do so as reliable statistical data is particularly hard to come by. Most sources agree that statistics can only give us a very rough overview of this complex phenomenon. The reasons are partly political as authorities of receiving societies tend to minimize, while those of labour-force-exporting countries tend to overestimate, the dimensions of migration. Another reason is that only migrants using official channels can be counted and that the number of illegal migrants, though certainly very high, is much more difficult to ascertain. Varying methods and differing statistical operations make comparability tricky and are another source of errors. Therefore I restrict myself to referring to Owen (1985; cited in Seccombe 1988: 199) who estimated that there were around 5.7 million[2] migrants in the Arab world in the early 1980s. The magnitude of this movement can likewise be derived from the exorbitant percentage of foreigners in the national work-force of these countries. In some states, like Qatar or the United Arab Emirates, foreigners exceed the nationals by far. Kuwait has been another salient example in this respect as Kuwaitis formed only 23.4 per cent of their own labour force in 1965, with the number decreasing still further to 18.9 per cent by 1985 (Shah and Al-Qudsi 1989: 32).

One of the Arab countries from which out-migration in the wake of the oil boom has been particularly important is Egypt. Migration, as a matter of fact, is not a new phenomenon in this country but actually has a century-old history; however, the major turning point came in 1973 when the number of out-migrating persons sharply increased. Before this watershed, out-migration had existed but had been heavily restricted by the Nasser government.[3] According to one assessment (Roy 1991: 554) the government saw in its labour force an asset too important for a country at war to do without. However, after 1967 out-migration was gradually facilitated and bilateral agreements were concluded with other Arab governments. Consequently, at that time out-migration largely took the form of government secondment or contracting for a fixed period. Accordingly, those who left for work abroad were qualified and took positions as teachers or professionals in other allied Arab countries. In the atmosphere of Arab nationalism reigning under Nasser it is no wonder that the sending out of this trained work-force was connected to the ideology of Arab unity and the mission to assist these nations in their plight for development (Sell 1987; see also Amin and Awny 1985: 31–3, and Dessouki 1982). It is therefore not surprising that migration across national boundaries during that period was an urban, rather than a rural, phenomenon.

The Egyptian government under Nasser's successor Anwar as-Sadat realized that the oil boom of the early 1970s constituted an important chance to alleviate one of its major internal problems: unemployment. It is no wonder, then, that the restrictive issuing of exit visas changed with Sadat's advent to power. Migration policy was radically modified and the export of labour power from that time onwards was largely encouraged. In 1971, migration, as a right of every citizen, was incorporated into the Constitution. Consequently, several administrative units were established to monitor and to facilitate out-migration. From the mid-1970s onwards it quickly reached new highs. Since that time Egypt has become a source of labour not only to the prosperous oil economies of the Gulf but also to almost every Arab country. International labour migration is thus closely linked to *infitah*, the economic and political opening Anwar as-Sadat initiated after the October War in 1973. *Infitah* was meant to be an economic liberalization – an 'opening' of Egypt for foreign capital and foreign commodities – but it was in fact migration which was to become economically, socially and also culturally its most significant manifestation (Said 1989: 24).

Again, most authorities agree that the validity of information available from statistical data is suspect. Therefore, only a tentative assessment of the numerical aspect of out-migration will be given here. Abdel Fadhil (1979; cited in Amin and Awny 1985: 1) estimated the number of Egyptian migrants for 1965 at 100,000. Birks *et al.* (1983: 118) indicated 160,000 Egyptians working abroad in 1973. In 1975 their number had risen to about 370,000 working in the Arab world out of a total of around 655,000 persons employed outside Egypt (Birks and Sinclair 1980). In 1980 their number had gone up to 800,000 workers (Birks *et al.* 1983: 114). In 1985, the year which is generally understood as the peak in out-migration, the number of new exits gradually declined; official sources estimated that there were 1.79 million workers abroad out of a population of 48 million (CAPMAS 1985: 1, 6; cited in Sell 1987: 30). In the same year, however, another source (Miller 1985) reported 2.5 to 3.5 million workers abroad. According to this high estimate as much as 10 per cent of the Egyptian labour force had a job in a foreign country. Boustani and Fargues (1990: 126) indicate an equally high rate, with 11 per cent of Egypt's active population working outside Egypt. Recently, Fergany (1992) spoke of over 700,000 Egyptians having returned from Iraq, Jordan and Kuwait in the wake of the Gulf crisis.

It is interesting to note the male bias in all publications: international labour migration has always been implicitly understood to be a male issue. That there is also a substantial number of Egyptian women working in the Gulf states as professionals or unqualified workers is only very slowly coming to the general awareness. Bendiab (1991: 118), quoting Sylvain (1981: 34[4]), puts the percentage of females among Egyptian migrants in the Gulf region at 30 per cent as early as 1980. As she states, many of them work in modern feminine professions: as teachers, physicians and nurses. However, in my own experience, in the village context out-migration is in fact still an entirely male phenomenon. To date, only *urban* Egyptian women migrate to foreign countries.

Scholars turning their interest to Egyptian labour migration initially described it as a largely urban phenomenon (Amin and Awny 1985: 200), but migration is to be observed in the rural context as well, and it was social scientists engaged in village studies who first pointed out this fact. And, interestingly enough, the Egyptian migrant is nearly always represented as the *fallah*, the peasant, dressed in his traditional costume – the *gallabeyya*.[5]

Though also widely differing with regard to the numerical impact of out-migration, scholars agree that migration in the 1980s was of tremendous importance in the Egyptian countryside, too. Here again are some statistics. Hopkins (1988: 136) and Commander and Hadhoud (1986: 172) mentioned around 10 per cent of migrant households in the villages they surveyed; Adams (1987), who worked in the Upper Egyptian village of Diblah, put the percentage of all households reporting a member abroad at 33.9 per cent. Hammam (1981: 10), basing her study on the economically active population, stated that in some regions of Upper Egypt, such as Beit Alam and Haraga, out-migration ranged between 50 and 75 per cent.[6]

All this shows that labour migration since the mid-1970s must, in fact, be of crucial significance to the Egyptian government which is confronted with the predicament of a high population growth and the task of creating new jobs. But the migrant workers' remittances also count among the state's major sources of foreign income. According to Bassiouni:

> the Egyptian economy depends on five sources of revenues: (a) expatriate remittances ($3 billion); (b) oil revenues, ($700 million); (c) Suez Canal revenues, ($500 million); (d) US aid, ($1 billion in non-military and $1.3 billion in military grants); and (e) tourism (which until 1986 was estimated at $700 million, but in 1988 was estimated to reach $2 billion).
>
> (Bassiouni 1988/9: 76)

Given the structural inadequacies of Egypt's labour market it is no wonder that in their private estimations Egyptians also attribute extremely high importance to migration. Everybody knows migrants, belonging to his/her family or friends or living in the neighbourhood. One also scarcely meets an Egyptian who has not at one time considered migration, even if he/she has not actually left the native community for work. Many think that today there is no household in Egypt which is not, in some way or other, directly or indirectly, touched by migration. I suppose that many foreign visitors to Egypt, especially those who come from the rich economies in the west, have in fact experienced this themselves, as one is frequently asked about the possibilities to get a work contract in the respective country. Migration as such is an important consideration for Egyptians, and they have a distinct priority list of migration countries on the top of which range such western countries as Germany or, of course, the US. This priority list is then followed by the rich Arab Gulf states such as Saudi Arabia, Kuwait or Qatar. Further, such countries as Iraq, Jordan or Libya are considered. At the very bottom of this wish-list ranges internal migration: such as going to work in one of the mushrooming tourist resorts in South and North Sinai, in the industrial zones of Suez, in the new desert towns, or in the so-called, notorious 'informal sector' in Cairo.

One of my initial arguments had been that migration today very often takes the form of a spatial movement of labour power within a globally organized system, and, emanating from so-called under-developed regions it is directed towards the world's centres of capital, production and services. The underlying motivation for this specification is my interest in going beyond a mere description of the confusing multitude of global migratory flows and to reach an understanding of the very functioning of such processes. However, rather than searching for a new grand paradigm which fits all these empirical emanations of migration, I want to restrict the analysis to the aforementioned subtype. By concentrating on the migration of Egyptian peasants to the Gulf states I intend to investigate the relevance of labour migration from subsistence-oriented contexts (what is often called the rural periphery of this world) to such centres where capitalist forms of production, i.e. wage labour, are prevailing. In this vein I have eventually focussed on the Egyptian countryside. This immediately calls for further elaboration and a more precise definition of my object of research.

Generally speaking, agricultural production in Egypt – the Nasserist land reforms and the politics of *infitah*, the political and economic 'opening' notwithstanding – is still dominated by small peasant holdings. Such smallholdings are often defined as comprising up to five *faddan* (a *faddan* equals 0.42 ha) and are usually estimated to guarantee basic subsistence of the household. Yet with regard to Kafr al-ʿIshra, the Egyptian Delta village this book is a case-study of, this definition is nearly meaningless, as those having access to more than two, or even to one *faddan* are already regarded as *mabsutin* (well-to-do). This opinion adequately mirrors the overall trend in Egypt's rural areas as farm sizes have been steadily declining since the beginning of the 1960s (Richards and Martin 1983: 4). Smallholdings (defined here as under 5 *faddan*) represented 78.5 per cent of the total number of holdings in 1950 and increased to 84 per cent in 1961 (Abdel Fadhil 1975: 14) and further went up to over 95 per cent in 1984 (The Middle East and North Africa 1989: 389). In 1984 the average size of these farms amounted to 0.9 *faddan* (0.37 ha) (Statistisches Bundesamt 1988: 43). In this respect one also has to notice that new alternative employment opportunities did not fundamentally change the economic situation of these households. So, Commander observed, even at the end of the 1980s

> between a quarter and a third of small farm households generate incomes on or below an estimated poverty line. Thus, the major changes witnessed over the last decade in the range of employment opportunities open to male members of agricultural households, have not necessarily been associated with a substantial and sustained improvement in income.
> (Commander 1987: 287)

Another aspect which has to be taken into consideration is that many families do not have access to land at all, or if so, only in a restricted way. I am therefore dealing in this book with small landholding peasant households plus the poor and landless who, though without permanent access to land, also derive their income mainly from agriculture.

This book, then, deals with these small peasant households' struggle to feed their families and to make ends meet under economically very harsh conditions. One of the principal strategies employed to reach this aim has been migration to the Gulf economies for work and this will therefore constitute the major focus of this study. It deals with these people's preparations to send a family member abroad for work, with their lives during the absence of the migrant, with the hopes they pin on migration, with the reintegration of the migrant member – in short with every perceivable aspect of migration in the lives of the villagers.

Before commencing, however, a theoretical framework is necessary. In Chapter 2 I shall have a critical look at what I found to be prevailing theoretical concepts of international labour migration in general and about migration to the Gulf in particular. There exists a vast body of literature about migration which is hardly possible to study in its entirety. Any discussion must therefore be selective and correspond to a specific framework of reference, and I do so here as both a development sociologist and as an anthropologist. As such, I start off by describing peasant migration within the overall framework of market integration – that is, the capitalist world market. Given this context I touch upon the organization of agricultural production in nineteenth-century Egypt, and further elaborate on similar present-day forms of Egyptian migration. Pointing to the relevance of Asian migration to the Gulf states further strengthens my argument – namely, that of the historicity and flexibility of globally organized flows of labour, the context within which Egyptian peasant migration has to be analysed.

I have already characterized labour migration as one crucial constituent aspect of the world system which can be conceptualized as being constituted out of a multitude of global flows. Another kind of global flow, which seems extremely important in this context but which has until now been largely overlooked, is of a cultural nature. Migrants do not only seek employment, refuge, a better life but they also function as symbols, they also make sense of the world around them, they and their families take a stand *vis-à-vis* other kinds of cultural flows. As with other aspects of migration I am not so much concerned with its circulation, its magnitude and its directions, but with the local level, with the ways people produce, reproduce, transform – in short how they make sense of such cultural flows. This is the theme of Chapter 3. I shall deal with my subject first of all from a theoretical level. The second part approaches the subject from a different angle by providing a narrative about my experiences when coming to Kafr al-ʿIshra, the village I did research in, for the first time.

While market integration up to this point has been viewed predominantly from a macro perspective, Chapter 4 deals with the implications of this system on the micro level of the village community. I shall sketch the historical genesis of the village and describe its present-day outer appearance. Further, in order to provide an understanding of a Third World village in the contemporary capitalist world market I shall deal with power structures and the main economic activities pursued in the village.

Chapter 5 is dedicated to the main characteristics of out-migration from the village community. I shall focus more closely on the subject by introducing a case-study of two particular streets in a poor quarter of the village. My special interest will be on the economic activities the households of these two streets engage in. I shall demonstrate the significant relevance of labour migration as one form of production besides others these small peasant households pursue in order to secure survival.

Departing from the fact that labour migration at the end of the twentieth century is not forceful, Chapter 6 explores the issue of migration as an economic and sociocultural strategy of the peasant households as well as of their individual members. I shall proceed by discussing why these peasants leave their land and turn into migrants and what they do with the money earned abroad. Prior to remittances spending, of course, are the practical aspects of preparation and the realization of the household's migration project.

As these peasant households are always short of monetary funds, how do they provide the cash for air fares and passports and how does the peasant family cope with the household head's absence? These questions are addressed in Chapter 7.

The concluding Chapter 8 deals with the effects of the Gulf crisis of 1990/91 on the village community. Many migrants returned as one immediate outcome of the crisis. But what happened to the migrants and what relevance has migration today for these small peasant households? Coming back to one of my main theoretical interests – namely, the flexibility with which the peasant household's labour force has been put to profit by external forces since the very beginnings of accelerated market integration – I shall explore the economic activities these small peasant households engage in during a time of severely restricted access to external labour markets. I finally take up again the issue of the interrelationships between different kinds of global flows, reaching the conclusion that the village is the product of the intersection of these flows, constantly changing as the global flows fluctuate in direction, intensity and content.

ON METHODS

This book is a case-study of an Egyptian Delta village which I have chosen to call here Kafr al-ʿIshra. The village is situated near the district town of Sinbillawein in Daqahliyya province. It was chosen predominantly because of its size and its location. As there are only 287 households, getting familiar with its community seemed to be easier than in a village with several thousand households, which is no rarity in Egypt. Also, fieldwork for a single western female

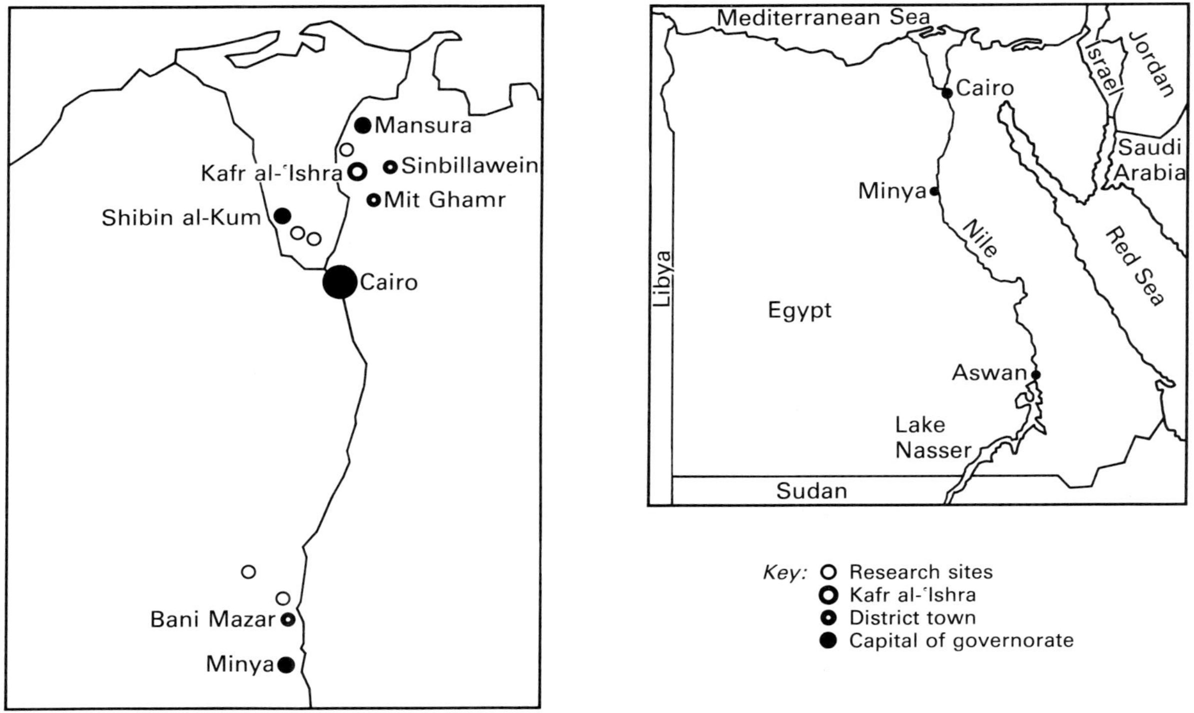

Map 1.1 Geographical location of research sites

researcher was deemed to be easier in Lower than in Upper Egypt. The rising importance of Islamic fundamentalism, widely spread in some regions of Upper Egypt, added to the choice of this small community as my main research site. My field research[7] about Egyptian peasant migration stretched over the months from March to October 1987, and I stayed in Kafr al-ʿIshra from June to September 1988 and again in June 1992.

Data used in this book basically consists of my own findings, derived predominantly from participant observation. But I also analyse statistical data and occasionally draw upon case-studies prepared by research assistants (see note 7). These statistics give interesting insights into the basic characteristics of the village community and its migrant households. The set of quantitative data permits the analysis of 'objective' structures and relations, and here I am especially referring to the functionality of migration from small, subsistence-oriented peasant households within the framework of market integration on a global scale. Yet, as it is only through the extensive use of qualitative material that we can grasp the emic view of the peasants and can give them their due as social actors within this system, most of the data on which this book is based has been arrived at by participant observation, informal interviews and conversations with the villagers.

During my fieldwork in Kafr al-ʿIshra I stayed in the household of the village headman. I was warmly accepted in his household as a guest and I soon felt like a daughter belonging to the family for the period of my stay. The headman's wife and his daughters-in-law became among my most important resource persons. They invited me to participate in all their activities in the household and to join them on their visits to neighbours and relatives in adjacent villages or to the market in the district town. I had a young assistant resident in the village who in the first weeks of my stay accompanied me and introduced me to the migrants' families. I established contacts with about 30 families and I decided to concentrate on about 20 of them. Most of these households were located in a poor Muslim (the community is predominantly Christian) quarter of the village. I usually paid my visits to these households on a daily basis for about two hours in the morning and another two hours in the afternoon. During the rest of the time I stayed with the women of 'my' household or joined the women at home, when they went to the fields, to other villages, or to the district town.

Most of my work was with the women, though talking with their men never was difficult or awkward for me – except sometimes with young, unmarried men. I avoided the latter – which, as I happened to learn – was observed and appreciated.

While I always had an issue on my mind, I never wanted to 'interview' the women, but I tried to keep discussions as 'natural' as possible. I tried to join them in what they were doing when I dropped in. Many times our conversations diverted from 'my' theme to touch other subjects, which often turned out to be of equal interest to me. 'Interviewing' with a questionnaire in my hand or a cassette recorder proved totally impossible for me. There are various reasons for this. First of all, data for many of my research interests could not be obtained in this way. Interviewing might perhaps lead to acceptable results when talking to a teacher or an employee in his or her office, but not when observing a woman pedlar who is selling her vegetables, haggling over the price of her tomatoes with a neighbour, wiping a toddler's nose, telling her young daughter to beware of the gas while frying potato chips for lunch, trying to get to know the latest village rumours, chasing the poultry away from her commodities – and me at her side with a cassette recorder wanting some information about this or that. There were many similar situations. And it was a matter of luck to catch one of them in a silent moment. It was at such a time that a woman might ask if I wished her to fetch a box containing silky materials, fancy, transparent nightdresses, bed covers brought back by her husband from Iraq or Saudi Arabia. She would comment on their usage and on how the fabrics and dresses could be kept for the wedding of a daughter. Most of the time I sat together with the women, following what they were doing, joining their conversations as well as possible, posing questions which made sense in the course of the discussion. I wrote down my field-notes in the long and hot siestas and in the evenings. These evenings also were filled with discussions with the two daughters-in-law of my host household. As they belonged to the most intimate of my relationships I related to them what I had observed during daytime, what I had been told, and often they helped me by elaborating and commenting.

I want to make another point regarding methodology. I was sometimes asked by the villagers what the use of such a study could possibly be, as there was no kind of development project linked to it. Given the extreme poverty of the villagers, my somehow '*l'art pour l'art*' project, undertaken to result in my acquisition of a doctoral degree, often irritated me to say the least. Yet finally I understood that my 'research-objects' and I were on fairly equal grounds. I often wondered who was asking the most questions – they or I. They, like me, had certain 'research interests' – their centring around agricultural production in Germany, religious practices, relationships between the sexes, food preparation. There even was something like a common theoretical background: while I was interested in 'modernity' they had a shared interest in the magnitude of differences between 'the developed world' and the 'underdeveloped village'. I also often noticed that the women told each other what I had said and made this a point of their discussions. They also used methods of 'participant observation' resulting in my feeling that I was only very seldom in a state of privacy. I have tried to do my best to answer their questions, hopefully with a similar amount of honesty – and concealment – as they did.

Finally, a comment on the transliteration of Arabic words. I am quite aware of its inadequateness for any orientalist or social scientist trained in Islamic sciences. However, for practical reasons, and in favour of readability for those not initiated into the secrets of western orientalists, I chose a very simplified form of transliteration.

2

EGYPT'S RURAL AREAS OVER THE PAST TWO CENTURIES

From *tarahil* to the rationalization of the international migrant business

As a socio-anthropologist attempting to come to terms, theoretically, with the issue of international labour migration from rural areas to the world's centres of production, services and capital, I have found prevailing migration paradigms of little analytical value for my own undertaking. Often these proved to be ahistorical, too economic, too static, and imprisoned in conceptions based on a kind of *homo migrans* circulating between push and pull poles of supply and demand. Research interests are nearly exclusively in the positive and negative (again mainly economic) impacts on the receiving and sending community, and with regard to the latter a focus on remittance spending is prevailing. And naturally, as an anthropologist, I have difficulties with the very nature of the overwhelming body of data used because it is mainly statistical. Dealing exclusively with figures allows the researcher to remain at his/her office desk, but with the resultant risk that he/she never becomes confronted with the realities of the persons being studied.

This chapter will therefore present my main points of critique of current migration concepts. I shall then proceed to sketch my own theoretical framework.

OUT-MIGRATION FROM EGYPT IN THE MIRROR OF SCHOLARLY TREATISES: A SKETCH OF BASIC THEORETICAL CONCEPTS

Trying to achieve an understanding of migration turned out to me to be an extremely unpleasant undertaking. Though there exists a vast body of migration literature, especially in the realm of economics, I had great difficulties in finding publications to inspire my own project. Much of it, I felt, was too abstract, and usually void of human beings. And the explanatory power of many quantitative studies remained unconvincing to me in spite of their making use of highly sophisticated statistical formula. My interest rose, however, when I recognized that there was a certain underlying paradigm inherent in much of these studies (or at least those I finally managed to read). Eventually I also realized that existing studies about migration in the Egyptian and generally Arab framework were usually inspired by the same model. There somehow seemed to exist something approaching a 'grand theory' applicable to a wide range of empirical migration settings. It is therefore because I became increasingly critical of this underlying paradigm the deeper I got involved with my subject that I here want to present some of the main characteristics of this model.

Labour migration according to this paradigm is conceptualized as a dual, a bipolar (e.g., rural-to-urban) movement, a circulation of migratory labour between one 'sending' and another 'receiving' pole. Consequently, reasons for this bipolar mobility of labour power are made out in terms of push and pull factors such as supply and demand. Research issues centre on the identification of negative push factors triggering off migration from the sending societal entity and, respectively, positive pull factors making out the attractiveness of the receiving society for the potential migrant. Migration, then, is seen as triggered off by the disequilibrium between countries rich in manpower and poor in capital and those which are characterized by the opposite features: poor in manpower but rich in capital. Further, the outcome of this process presented in the conclusion often seems to reflect a general orientation of the researcher. Those subscribing to one or another form of modernization or to neoliberal economic theory look forward to the eventual achievement of an equilibrium to the benefit of both poles. Their opponents subscribing to the dependency paradigm, on the contrary see the effects in a gradual emptying of the sender-society of its valuable and precious resources. Studies about the brain drain or about any other qualified labour force, or about the negative results of consumerism remigrants engage in where capital which could be used for some investing purpose is squandered, are typical cases in point here.

Here is a good example of this kind of approach, given by the secretary-general of the OECD on the occasion of the introduction to a conference on migration. He cited as the main causes of migration:

> First, there is the difference in demographic potential. Throughout history, empty and underpopulated territory has attracted the overflow from heavily populated areas. . . . The second difference in potential is economic. Migrants are looking for better jobs and higher living standards. . . . Lastly, there is the difference in political and cultural potential. A major reason for migrating has always been the search for a country where life is easier, where there is more freedom and the life-style and cultural climate are less oppressive, strict and confining.
> (Paye 1987: 9)

It is obvious that we here have one of the cases where the so-called exact sciences, in this case physics, are drawn upon to explain an empirical fact which is clearly situated within the realm of the humanities. Paye himself refers to this background in the following way when introducing his account:

> I vaguely recollect from my far-off and very elementary physics lessons at school that electric current flows all the more freely between two poles the greater the potential difference and the lower the resistance of the circuit. It is this somewhat far-fetched comparison that occurs to me when considering the causes of migratory flows.
> (Paye 1987: 9)

The circulation of migratory labour power between the sending and the receiving pole is another important aspect of this concept – the concentration is on movement, on the spatial mobility of these migrants. What is obscured are the very conditions which create this circulating 'element', the migrant. Rather, this migratory labour force is a somehow taken-for-granted constituent of the system and it seems that it only needs a certain combination of catalytic push and/or pull factors to activate the process. Here again the affiliation to some sort of grand theory explains the focus on specific push factors. While within the framework of dependency and the historical–structural school in general, poverty or unemployment caused by structural constraints are focused on, neoliberal tendencies bring the individual's interest in profit-maximization to the fore. According to this latter approach the individual chooses his/her working place where wages are highest, and if this decision involves spatial mobility he/she is turned into a migrant. Thus, the notorious *homo oeconomicus* of liberal economics has found its sub-species, the *homo migrans*. The *homo migrans* effortlessly circulates as a labour force which works at any given time, at any given task, and at any given place, once the process is triggered off. In these theoretical concepts the image of physical elements, which seemingly obey without question some natural law or structural force, arises before my inner eye. I do not deny the attractiveness of such a model, the lucidity and plainness of which seems so convincing at first sight. Yet it appears to me that explanations of this kind as to why people migrate do not go beyond the surface of the phenomenon. A perhaps somewhat polemical objection to this sort of 'profit maximizing conception' – that people turn into migrants because they want to earn more money – would then be: yes, without doubt, but why then does not every person of working age become a migrant as there will always be some place where one could earn more money? Similar questions could be posed to the structuralists.

Migration from Egypt to the Gulf states, prima facie, appears to be an excellent case in point for the described kind of paradigm: Egypt's population density is among the highest in the world and unemployment is rampant,[1] whereas the Gulf states are under-populated and thus lack an indigenous work-force but are rich in capital. I do not want to offer a detailed review on migration literature in this respect.[2] Rather, I want to exemplify my point by analysing some basic assumptions of what is admittedly one of the cruder representatives of the kind of argumentation outlined above. Choucri (1977), in an article called 'The New Migration in the Middle East: A Problem for Whom?', reproduces the above described stereotype of an inter-societal disequilibrium, by applying it to the Middle East, and especially to Egypt:

> The petroleum crisis of October 1973 has drawn dramatic attention to the difference among the countries of the Middle East in terms of size and wealth. Indeed, the most populous countries are the poorest in natural resources, and the richest countries are smallest in population. In addition, vast differences exist in the level of technological development and manpower characteristics. The largest countries in population size are the most developed in terms of overall knowledge and skills. For the smaller, but wealthier countries, manpower

> availability is the single most important constraint in economic development. Over the years, these differences have contributed to sharp patterns of migration across national boundaries and to the mobility of both skilled and unskilled labor.
>
> (Choucri 1977: 422)

There is another point which is highly disputable in the kind of paradigm presented above. This is the separate treatment of 'external' and 'internal' migration. Here, external migration focusses on migratory flows across national boundaries, i.e. between national entities. A major shortcoming of such an understanding is that hereby migrant sending and receiving societies are seen as homogeneous entities void of any internal differentiation or fragmentation. Rural–urban and socioeconomic cleavages and possible distinctive effects of migration on these sub-systems or sub-sectors remain outside the sphere of examination. Other forms of migration, for instance of the rural-to-urban sort, and the interrelationships and interdependencies between external and internal migration also go beyond the scope of the analysis. Also, the functionality of boundaries for the regulation of migration flows has only seldom been considered. In this respect it is interesting to recall Mitchell's argument:

> [The] convention of imagining countries as empirical objects is seldom recognized for what it is – a convention. The relations, forces, and movements that have shaped people's lives over the last several hundred years have never, in fact, been confined within the limits of nation-states, or respected their borders. The value of what people produce, the cost of what they consume, and the purchasing power of their currency depend on global relationships of exchange. Movements of people and cultural commodities form international flows of tourists, television programs, information, migrant workers, refugees, technologies and fashions. The strictly 'national' identity of a population, an economy, a language or a culture is an entity that has to be continually reinvented against the force of these transnational relations and movements.
>
> The apparent concreteness of a modern nation state like Egypt, its appearance as a discrete object, is the result of recent methods of organizing social practice and representing it . . .
>
> (Mitchell 1991: 28)

Turning to another example, I here want to refer to an author who ranks prominently among Arab scholars working on migration to the Gulf region. Ibrahim's publication on what he termed the 'new Arab social order' was considered to be one of the most important treatises of migration in the Arab region when it first appeared. Ibrahim (1982a, 1982b) in a very structural manner focuses on circulation between Arab national entities. For him, socioeconomic linkages in the Arab world are 'dramatically manifested in the flow of manpower and money (in opposite directions) cross the borders of the Arab states' (Ibrahim 1982a: 62). In his endeavour to substantiate his concept of the 'new Arab social order' he then works out a typology within which he locates all the Arab states, a system which has come into being as the result of oil revenues and its multitudinous 'spill-overs'. Ibrahim thus categorized the Arab countries into a hierarchy of four groups: the rich, the well-to-do, the middle, and the poor (Ibrahim 1982a: 58). This 'new Arab social order' is kept together by the already known processes of circulation in opposite directions. Of prominent importance, of course, is the flow of manpower. A step forward in Ibrahim's modes is his emphasis on the integration and interdependency of different kinds of flows within one system: one instance is trade – e.g., raw materials vs food, arms, manufactured goods (Ibrahim 1982b: 163); transfer of capital (in two directions), of technology and managerial know-how (one-way), as well as aid (financial, technical, commodities or managerial know-how) (Ibrahim 1982b: 164–5). And yet, the cheapness of the labour force is explained by an 'overpopulation' or a 'surplus population' which makes labour power available in abundance in one country while it is scarce and therefore expensive on another national market. Here again the focus is on circulation triggered off by an imbalance in the market. Ibrahim's model was a step forward as he enlarged the concept of bipolarity to encompass all Arab states and furthermore made it clear that 'the Arab world is fully integrated in the capitalist economic system' (Ibrahim 1982b: 164). But his definition, it seems to me, only went half way, as migration was still conceptualized in national categories and as a circulatory process between two poles (if within this larger framework). Further, his concept of a distinct Arab system is somehow reminiscent of some kind of super-nation existing within a global context. Boundaries not only exist between the Arab states but there also is something like a super-nation embracing all the Arab states. Still, interdependencies between this

Arab entity and the rest of the world remain unclear and are faded out, not being considered worthy of any particular scrutiny.

Conceptual considerations suggest the abandoning of models focusing on national entities carved out of the rest of the world in segregated, disconnected systems in favour of an encompassing and dynamic historical framework. The Middle East is no 'closed system', as Choucri (1977: 422)[3] once maintained in his description of the Arab 'super-nation', and even Ibrahim's notion of the 'New Arab Social Order' is highly disputable. Carving out one particular region of the world without taking into account the multitudinous interdependencies and flows – that is, without considering that market integration on a global scale and its forces at work on the local level of these societies are of crucial importance – means that any analysis must remain a serious misconception. On an empirical level, it is at least the large-scale immigration of South Asians especially, and the increasing replacement of Arabs by Asians, which falsifies such conceptions. The ongoing rumours and discussions about which nationality of migrants are to replace the undesired Palestinian labour force in post-war Kuwait – Turks, Egyptians, Asians, Kuwaitis – is just another case in point. When looking for a macro-structural framework within which migration can be understood we must therefore take into account the fact that today not only is there no region which has not been subjected to market integration but also that there is a considerable dynamic element inherent in this process.

This brings us back to the notion of circulation which implicitly takes the very existence of migrants for granted. The question of how these migrants come into being, how the migratory labour force is 'produced', and of the consequences of migration for the networks they originate from is not posed here. This 'physical' conception suggests that migrants are isolated and detached elements, and that the location they leave from is not influenced or concerned by their departure and is not relevant for an understanding of migration. Yet, as migrants are human beings they are integrated into a social world and into networks. They belong to a household, they have a family, relatives, a neighbourhood, friends. Within these networks their practical everyday lives are embedded. It is evident that such a major event as a person leaving for a considerable period of time or his/her return must have its influence on fellow persons belonging to these networks. Furthermore, it is to be expected that they also take a stand on the matter of migration. The closer these relationships the greater the impact of migration must be: the everyday lives of members of a nuclear family without doubt change dramatically due to the migration of a parent. My argument, then, is that migration cannot be the concern of the migrant alone but must inevitably be that of all family members. The relevance of migration for the practical everyday lives of these persons, moreover the existence of such networks, has until now remained a veritable black spot in the bulk of studies on migration. But without these networks migration would hardly be possible. This is all the more the case with regard to the specific form of migration under scrutiny here: that from rural areas and from agricultural households. These effects of migration on the sender community have not been examined in most migration studies, which are usually concerned with exclusively economic issues (i.e., the spending of remittances and its implications for the national economy). But, especially in a more traditional society, these effects cannot be measured exclusively by counting how much of their remittances migrants spend on longlife consumer durables, how many new houses they build and how much is invested in non-agricultural, income-creating projects. Even the migration decision itself must have repercussions on a rural household as funds must be diverted for the preparation of the journey, and the migrant's labour force must somehow be replaced. Upon his return he expects to be reintegrated, not only into his household but also into his neighbourhood and his community. How does this affect the women staying behind and, of similar importance, what attitude do they develop towards migration? These themes have seldom been mentioned, let alone dealt with in a larger research project. Only a small number of publications, mostly articles, address themselves to the consequences of male migration on women left behind and on their status and roles.[4]

In short, one can expect that migration is relevant to the social networks migrants originate from. Instead of departing from some *homo migrans* concept, one has to consider the social networks he/she is integrated in as without these structures migration would be hardly possible at all. In other words, if the existence of a *homo migrans* cannot simply be taken for granted we can argue that as migrants must have somehow come into being – either from the ranks of peasants, the unemployed, small shopkeepers, agricultural wage labourers, and fathers and husbands – we have to include these 'migrant producing contexts' into the analysis.

MIGRATION AND THE WORLD MARKET OF LABOUR: A THEORETICAL OUTLINE

How, then, to come to terms with the phenomenon of the large-scale migration of Egyptian peasants? If dualistic push and pull concepts are too restricted and are, moreover, ahistorical and hence of unsatisfying explanatory power, what alternative frameworks can be taken into consideration for a more convincing explanation? It is these questions I will address myself to now. The problem has a macro- as well as a micro-aspect. In other words: answers must be sought for by turning to the mechanisms of the world economy as well as to those of the peasant migrant's household. While I will deal with the latter later on (see pp. 101–40), this chapter will be dedicated predominantly to the macro level. I shall sketch a theoretical framework and refer to empirical cases, drawn from an earlier phase of world market integration and from the end of the twentieth century, to substantiate my arguments.

By describing the phenomenon of mass migration we could easily establish different analytical categories of migrants (such as refugees, the brain drain, internal and external, to name only some of them), but instead of engaging in such an endeavour I suggest focusing on migration as a process of spatial mobility of labour. Thus, a person may have to travel varying distances to reach his/her working place, which may not allow for returning home after a working day. In other words, whenever a person does not return to what he/she considers home after a working day due to having to travel long distances, and if this circumstance persists for a lengthy period, then some definition of migration comes up. More interesting in this respect than concentrating on the very conditions of movement (i.e., its periodization or the poles between which the movement takes place) is to take into consideration labour as a commodity and its availability on a globally integrated market. This then brings us to a historical perspective from which to understand labour as one cornerstone of the modern world system (Wallerstein 1974). This term refers to a conception which takes our world as having ceased to consist of sealed-off societal and economic entities, but as one globally articulated system where all sectors are interdependent and connected by various forms of exchange and flows. According to Wallerstein, this process started in Europe during what he has termed the 'long sixteenth century' (which he specifies as having lasted from 1450–1640), and is the process of rising capitalism and its evolution into one global capitalist economy. This process is commonly understood as a reorganization and adaptation of local economies to the needs of the world market. Very roughly speaking, Wallersteinians draw a distinction between the 'centre' and the 'periphery', allowing for intermediate stages of 'semi-peripheral' societal entities. Within this frame the stress is usually put on the peripheral economies being transformed to meet the needs of the centre: the extraction of raw materials and the shifting of agricultural production from produce which is consumed by the producers to cash-crops for external markets. In turn, advanced, mass-produced industrial commodities find new markets in these underdeveloped – or underdeveloping, for that matter – countries. Wallerstein further emphasized the fact that the 'world economy' is characterized by one single division of labour, structured within one global frame. Rising capitalism, therefore, is a process whereby production is increasingly globally organized.

It is of crucial importance to understand that within this frame the control of labour takes manifold forms. In fact, wage labour, which is generally associated with capitalism, is only one, even privileged possible mode to recruit labour. Besides this, unremunerated forms of labour have been historically more important. Here, slavery, corvée, forced cash-crop production, among others, play a vital role. This then leads us to the fact that commodities are not only the producers' agricultural or manufactured products – the *materialized* labour – but that these also comprise the producers as a product themselves (i.e., the *living* labour, to put it in Marxist terms – see also Potts 1990). Labour power itself is an important commodity. Such a perspective then allows us to deal with migration within the framework of commoditization. Spatially mobile (living) labour constitutes just one distinct form of the flows of locally produced commodities to globally organized markets. At this level of analysis national boundaries and the specific distance separating the labour power's *loci* of production and valuation become of secondary importance.

Migratory labour is thus one important element of a world market of labour, of a globally organized system of labour supply: this gives us a much clearer idea of the central issue of migration than any sophisticated categorization of different types of migration such as brain drain, rural to urban migration, national or international, legal or illegal, environmental, coercive or refugee migration (see also Portes and Walton 1981: 25, Appleyard 1989). Labour migration in the current historical period has to be seen instead as one factor 'articulating' with the general trends of the internationalization of production (Cohen 1987, Portes and Walton 1981, and Sassen 1988). Migration is not the only element of this system. Fröbel *et al.* (1977) have pointed to another important constituent factor in their description of what they termed the 'new international division of labour'. Their focus is on the emerging of a 'world market for production sites', of 'free production zones' and of world market factories (Fröbel *et al.* 1977: 30–1), a qualitatively new development in the world economy since the 1960s. They spoke of a 'veritable worldwide industrial reserve army'. However, they did not explicitly deal with the simultaneous transfer of living labour (see also, for a critique, Cohen 1987: ch. 7). It is Sassen who was among the first to stress the relevance of (living) labour as one important internationally circulated commodity among others. To her, the 'Internationalization of the economy resulted in the formation of a transnational space within which the circulation of workers can be regarded as one of several flows, including capital, goods, services, information and values' (Sassen 1988: 3).

These migration flows, according to her, reflect the rising of 'global cities', of Export Processing Zones and world market factories as a consequence of

> the shift in the location of rapid industrial growth from the old industrial centers to peripheral areas, notably the new zones in Southeast Asia, the Caribbean Basin and Mexico–United States border, as well as the vast industrialization programs in OPEC members.
>
> (Sassen–Koob 1983: 176)

Migration, then, in spite of its recently exploding numerical relevance, is not only a contemporary phenomenon but is also closely linked to the formation of the capitalist world economy. Though migration definitely occurred in earlier phases of world history, and probably has always been an accompanying factor of human society, it is crucial to capitalist market integration as one element of the globally organized availability of labour power. According to Potts (1988: 17–18), migration for work covering large distances can be dated back to the end of the fifteenth century. She locates its beginnings in the enslavement of Red Indians in the wake of the conquest of America. She also points to different forms of forced labour and forced migration in Latin America, Asia and Africa, to the slave trade which exploited African men and women, to the coolie system, and to forced labour in Europe in the twentieth century. Slavery is perhaps the most spectacular case of a highly mobile labour force, and the transatlantic slave trade probably has been the largest forced migratory movement in world history. Black slaves were coerced to work especially in the Caribbean, Brazil and the Southern States of the USA (Potts 1988: 55). From 1451 until 1870 about 9.5 million Africans were enslaved and brought to the European colonies, and another 50,000 to the European metropolises (Wirz 1984: 35–6).

However, we need not go as far as the Americas to find African slaves, we can stick to the empirical case of most interest to us here, as 'every important town in nineteenth century Egypt had a slave market' (Walz 1985: 137; see also Baer 1969), fuelled by slaves especially from Sudan. Though slavery was officially prohibited in 1855 the decree had little effect and clandestine trade continued until the beginning of this century (Walz 1985: 138). Slaves were also working in agriculture at that time. Baer (1969: 165) points to the fact that agricultural slavery was not uncommon and that black slaves were used on the Khedives' huge plantations and also by peasants in Upper Egypt. Baer (1969) goes on to remark that 'in Lower Egypt agricultural slavery was the temporary result of the sudden prosperity in the 1860s which led to the acquisition of new land and to a tremendous expansion of agriculture'. There were other forms of coercive mobile labour in Egypt and we can easily observe its coincidence with accelerated market integration. Egypt had for centuries been an important provider of cereals to European urban areas, but it was from the beginnings of the last century onwards that the land and the work force of its people was to an increasing extent subjected to the interests of an international market. During Muhammad Ali's reign (1805–49) this gearing to external needs was tremendously sped up, and he could only implement his huge modernization programme through the extensive use of corvée labour in infrastructure and agriculture which often implied migration over considerable distances. Tucker tells us that

> As many as 400,000 peasants could be called to work each year for an average period of four months. The Mahmudiyya canal was built by the forced labour of approximately 315,000 peasants who were supplied by village *shaykhs* from seven

> different provinces. Peasants had to supply their own food and shelter during an enforced period of absence from home. Anywhere from 12,000 to 23,000 men, women and children, laboring without proper food, lodging, or tools, died on the project in a ten month period in 1819.
>
> Tucker (1986: 27)

Another kind of corvée labour extorted from peasant households at that time was for military recruitment:

> Muhammad Ali's political ambitions dictated the fielding of a sizeable army. After various attempts to raise an army of slave and Sudanese recruits had failed, he started a massive program of peasant recruitment. By 1830, after he had directed provincial and village officials to supply set quotas of men, the Egyptian standing army and navy numbered from 54,000 to upwards of 75,000 troops, supplemented by from 15,000 to 24,000 Bedouin irregulars. . . . When the *hatt-i sherif* of 1841 insisted on the reduction of the Egyptian army to 18,000 men, the peasant-soldiers were not demobilized but rather sent to work on irrigation projects or landed estates. . . . If not caught in the net of corvée labor for irrigation projects or military service, the *fallah* was still subject to forced labor in state industries and mines. Muhammad Ali's industrialization scheme created some 100,000 factory workers, but it is not clear how many of these were *fallahin* as opposed to displaced craftsmen.
>
> (Tucker 1986: 27–8)

These forms of forced migration in turn resulted in another kind of spatial mobility as thousands of peasants abandoned their land and fled to the cities, hid in the desert or travelled as far as Palestine and Syria to evade the heavy tax burden, conscription and forced labour (Rivlin 1961: 116, 204). Some of them stayed for good as Egyptian communities still in existence in Israel remind us.[5]

COMBINING PERMANENT AND MIGRATORY LABOUR POWER, SUBSISTENCE AND CASH CROP PRODUCTION: THE CASE OF THE *'IZBA*

In the preceding section I have argued that (labour) migration must be viewed as one constituent element of a globally organized labour market. From this (macro) perspective I have dealt with the spatial mobility of the labour power provided by slaves, corvée labourers, soldiers and wage labourers. In what follows I shall focus more closely on the organization of production of this 'living' and 'materialized' labour. I shall eventually argue that production in Egypt's rural areas of the outgoing twentieth century still shares essential characteristics with that of the last century.

Production, in the framework of the capitalist world market, is usually organized by connecting different forms of labour. Wage labour is far from being the only, and presumably is not even the prevailing, form. Rather than being remnants of past historical periods, slavery, corvée, service tenancy, unremunerated housework, subsistence production are all at times combined with wage labour in this system. The contemporary world market of labour is characterized furthermore by its enormous, steadily increasing flexibility and sophistication with which different forms of labour are articulated. Yet, many aspects of this resilience have already been in existence for the past two centuries. This is definitely the case with the *ʿizba*[6] in rural Egypt which I shall therefore deal with now. The *ʿizba* was a form of cash-crop production on large plantations, practised mainly in lower Egypt.[7]

The historical roots of the *ʿizba* system go back to the first half of the nineteenth century, when, under the reign of Muhammad Ali and his successors, major shifts in agricultural production were introduced. Generally speaking, these transformations were linked to the increasing orientation of agricultural production towards external (i.e., European) markets. This process was accompanied by a tremendous, steadily intensifying debt crisis in the Egyptian state, a solution to which was sought in ever-increasing taxation enforced upon the peasantry. Many peasant households lost their land as they were no longer able to pay what they owed.[8] The government's increasing need for cash, on the other hand, led to this legal securing of private landed property, a gradual process which had its landmarks in the legislations of 1836 and 1842 and culminated in 1858 when *ʿushuriyya* land, which comprised a large part of the cultivable soil, came to be full private property. It is especially in the Delta that this legal sanctioning of private property led to the establishment of large estates oriented basically towards the production of cotton for export.[9] The labour force on these estates was recruited from the impoverished and indebted peasants who had lost their land due to exorbitant taxation and the privatization of land. But also those who had fled their villages escaping from recruitment for forced labour or military conscription often sought refuge on the *ʿizba-s* as these had been initially exempted from corvée. This was the case with those estates which had been established on formerly uncultivated land (Baer 1962: 32–3, Wallace 1883) where the new landowner was rewarded for his reclaiming efforts with the exclusion of his labourers from corvée and military service, and was thus an impetus for the peasantry to settle on these newly established estates (Rivlin 1961: 62). According to Ramzi (1954: 5) viceroy Ismail (1863–79) even temporarily stopped the establishment of such *ʿizba-s* because of the extremely large number of peasant families seeking rescue and looking for protection against corvée and military recruitment.

Owen (1981b: 146) suggests that a lack of well-developed markets in agricultural labour resulted in the landowners' employment of former peasants, now turned into service tenants. These provided a certain regular amount of labour on the estates in return for a small plot for the cultivation of subsistence crops – a system which must have appeared on the royal estates as early as the 1840s. The *ʿizba* system, then, seems to have been especially implemented on such very large estates whereas smaller ones were operated in some kind of share cropping arrangement; but mixtures of both systems were also in use (Richards 1982: 68). The *taʿmaliyya*, as these service tenants, the principal permanent labour force on these estates, were called (see Nahas 1901: 140–4), were not fully proletarianized, as we have already seen: the basis of the *ʿizba* system was the exchange of labour for a dwelling and a plot of land. However, they had no claim to the land except for usufruct rights, as the land remained the property of the landlord. They had to spend a considerable time (Richards 1982: 62, speaks of 25 days per month) working on the plantation of the landlord. Subsistence production, that is production which is determined to enter the *taʿmaliyya* households' consumption cycle, was thus predominantly an unremunerated task for women.

Besides this basic feature of the exchange of a dwelling and a subsistence plot against labour service, the internal organization of the household's labour process and arrangements between the estate owner and the *taʿmaliyya* varied widely. The labour service of the peasant family was usually remunerated in cash, but it could also be discharged on a share-cropping basis. Leasing conditions of the subsistence plot also varied from cash rents to diverse crop-sharing arrangements (Nahas 1901: 134–40; Owen 1986: 76–7; see also, Tucker 1986: 35). The balance between direct exploitation of the land under the direction of the landowner or his deputy and leasing to the cultivators was equally adaptable to the needs of the landowner. On the one extreme he could farm the bulk of the land himself while leaving to his cultivators only a small plot for subsistence crops. On the other extreme nearly all the land could be leased except for a small portion. The advantages for the estate owner lay in the securing of a permanent and easily accessible male labour force, which could still be supplemented and augmented in times of need by that of the cultivators' wives and children (Tucker 1986: 43). These year-round service tenants were supplemented by a second kind of work-force – that of the *tarahil*: seasonal, usually migrant workers. *Tarahil* workers from Upper Egypt (see Nahas 1901: 145 and Richards 1978: 506) or the neighbouring villages left their home villages for a period of two to six months to work as daily labourers on these estates. Lambert (1943: 231–2) reports that *tarahil* labourers were housed in groups in tents, and that they never received foodstuff. Because of their poverty and in order to avoid having to buy from the estate the *tarahil* brought along the nutrition from their home villages and Lambert describes how '[O]ne can see them with large sacks, containing bread, onions and rice' (Lambert 1943: 231–2; my translation).

Inherent to the system is that this labour force was relatively easy to control and to discipline. Though nominally *taʿmaliyya* workers were free in the sense that they were not legally bound to the estate, this freedom rather served the opposite end – that of binding them to the estate – as they were always menaced by the threat of losing their status and expulsion (Nahas 1901). Being landless, staying on the plantation was the best chance they had as it provided them with usufruct rights on a plot of land, small as it may have been. But their existence on the *ʿizba* was always threatened by the availability of *tarahil* workers who did not even enjoy such usufruct rights and who had a job only for some months and thus must have been eager to take over the *taʿmaliyya* workers' place. The binding of the labour force to the estate was also secured by another effect: though service tenants were autonomous in the organization of their reproduction, the logic of the *ʿizba* system was not one of self-sustaining subsistence production but one of latent insufficiency, which necessitated a constant integration into commodity production and wage labour. Nahas (1901) gives an example of how peasants were *de facto* bound to the estate even when remunerated for their work. He shows how cotton was to be delivered to the estate owner who paid the cultivators in cash. However, in reality, due to their indebtedness, peasants usually did not receive anything:

> The cultivator, in the course of the year had to procure something to nourish, to dress, and to satisfy the needs of his family, sometimes to marry, to bury a parent etc; it was from the patron, that he demanded all that he needed to satisfy all his diverse needs. . . . Cereals and all advances in kind are in an arbitrary manner estimated by the patron for a price he esteemed as appropriate without control or possible protest.
>
> (Nahas 1901: 135; author's translation)

One rationale of the system thus lies in the tying up of its workers to the *ʿizba*: while providing for their own subsistence they were forced to seek integration into the market system, simply because their basis of subsistence production was not sufficient, moreover always endangered by the threat to be expelled.

One important characteristic of the *ʿizba* system of production was its high adaptability to the capitalist interests of the landlords who decided on what, when and how much the service tenants had to produce (Nahas 1901). The system was characterized by its high flexibility with which it could easily be adapted to the changing dictates of market forces. Flexibility was therefore secured by the combination of different forms of production. It would be wrong to understand it as an institution which combines a surviving, pre-capitalist form of production (the peasant family's autonomous organization of subsistence production) with modern, capitalist forms (wage labour on the estate). The *ʿizba* system was capitalist, even if its inner organization partly resembles non-capitalist forms of production. Capitalist interests of valuation determined the working process as a whole and subsistence production was to be maintained because of its usefulness, even indispensability, within the organization of production. The articulation of these different forms of production is the cornerstone of the *ʿizba* system of (market) production. It secured to the owners of large estates, interested in cash-crop production, the permanent availability of the necessary amount of cheap labour power which was guaranteed even if the worker himself was absent because of illness as other family members had to replace him (Lambert 1943: 225). An additional merit of the *ʿizba* system to the landowner was that his estate could be run as a single unit, thus minimizing costs of input while at the same time allocating different uses to different parts of the estate (Owen 1981b: 230).

Another major advantage for the landowner was the cheapness of the labour force which principally resulted from the fact that its (re)production remained excluded from the capitalist working process on the plantation. Both sorts of labour – that provided by the service tenants, the *taʿmaliyya*, and that of the seasonal migrants, the *tarahil* – were reproduced in their own households: while the service tenants for this end were provided with a small plot of land we have seen that the migrants brought with them foodstuffs from their own households. The division of the labour force into permanent and seasonal workers further minimized production costs as no surplus labour had to be paid during seasonal slack times. The *ʿizba* system obviously served the capitalist interests of the rural aristocracy, the absentee landlords and the state. Cynical as this may appear, it also had advantages for the landless cultivators in their constant plight to consolidate an 'autonomous' unit of production as it provided them with a small plot of land they were free to cultivate according to their own rationale.

With the Nasserist reforms of the 1950s and 1960s, the *ʿizba* as a unit of production ceased to exist as one effect of the land reforms was that large absentee estate owners vanished from the rural scene, part of their land being distributed to small peasant households and the former *taʿmaliyya*. Yet inequalities in the distribution of land and in the externalization of values produced inside the village remained. Stauth (1983a) has stressed that though the institution of the *ʿizba* is now an empirical phenomenon of the past, the very organization of the working process of cash–crop production and the ways it serves the socioeconomic elite in the village as well as external forces has largely remained intact until today. This is a very important issue in our context as well – that is, with regard to migration. Now as then, the small peasant household autonomously organizes subsistence production to a large extent and simultaneously engages in market production and wage labour. With labour migration to the Gulf the system has been enlarged tremendously geographically speaking. While I shall in later chapters return in more detail to the peasant household and its organization of production I shall, in what follows, show that *tarahil* are still an important factor in Egypt's economy and society.

TARAHIL AT THE END OF THE TWENTIETH CENTURY

Many aspects of the original *ʿizba* system of production are still discernible in contemporary Egypt. In the last century Egyptian peasant households produced labour for external markets as they still do today. Now as then, part of the peasant household's labour force is deployed in capitalistic enterprises even though often separated by large distances from places of production.

As a matter of fact, two decades of migration to the Gulf have brought this particular form of migration to the attention of scholars, while *tarahil* and other less spectacular internal forms of migration have often been overlooked. But even today *tarahil* labour exists (Messiri 1983, Toth 1991) and is of considerable significance; therefore I shall turn to contemporary *tarahil* labour now. These internal migrants work, for example, on the state's agricultural enterprises on newly reclaimed desert land. These new estates are again primarily geared to market production. An article by Sukkary-Stolba (1985) is of special interest here as she describes the different forms of farms and labour deployed on them. Not surprisingly we find the new settlers' wives engaging in subsistence production and becoming peasants. Even graduates' wives who probably never had been to a village prior to their settlement on the newly reclaimed land had to get used to the roles of a peasant woman (Sukkary-Stolba 1985: 186). Sukkary-Stolba states that those who finally manage to cope with the hardships and stay on the land 'had to learn slowly the skills that settlers have mastered for thousands of years'. A woman described her dilemma as follows:

> I went to the University for four years. Now I have to be a *fellaha* (female farmer). . . . I have to learn how to raise chickens . . . supervise the work crew . . . live with the flies . . . and above all speak like a *fellah*. . . . I do not know what my family would think of me when I would go to visit them in Cairo.
>
> (Sukkary-Stolba 1985: 186)

Here again we also find a significant amount of *tarahil* labour which is recruited by contractors and mostly consists of children and youths of both sexes from neighbouring governorates. These *tarahil* are also low-paid contract labourers hired for a distinct period of time. They are accommodated in dormitories for the length of their stay where they are also usually responsible for their food, which typically consists of what has been prepared for them by their wives, mothers and sisters back in the village.

While foreigners may only visit these land reclamation projects in rare cases it is another form of contemporary *tarahil* labour which can be detected by an attentive tourist on Cairo's mushrooming construction sites. These labourers, sometimes joined by their families, come and live in small makeshift huts erected on the premises of the working site. While the men work as labourers their women are occupied with housework (i.e. all the unseen chores needed for their husband's recreation after the long working hours). But sometimes one can also come across women working on the construction sites carrying stones or sand in large baskets on their heads. Those tourists leaving Cairo for a diving holiday in South Sinai virtually run into migration when they depart from the bus terminal in Abassiya/Cairo. Egyptian labourers heading for a working place in Jordan, Iraq or Saudi Arabia and going by bus and ship rather than by plane also leave from this terminal. It is indeed a strange scene bringing together Egyptian *fallahin*, dressed in their *gallabeyya* and spending their waiting time by sipping dark and sweet tea, and the travel-experienced 'backpacker' leisure generation in turn carrying with them what has come to be known as 'tourist water' – the characteristic plastic bottle of water which seems to have made mass tourism to underdeveloped countries possible. Both groups, though waiting in the same place, keep separate and draw unseen territorial boundaries around themselves which are only broken by furtive and surreptitious glances at each other. Arriving in the tourist resorts of Sharm al-Sheikh or Nuweiba one is again confronted with the products of the migration of the *tarahil* as it is certain that all the hotels built literally overnight in recent years arose under the hands of Egyptian peasants from Upper Egypt or the Delta.

When I go to Sinai I am always struck by the strange relationship between these distinctive migrant groups which, though all being engaged in spatial mobility and clearly dependent on each other, are yet so different with regard to their socioeconomic and cultural characteristics. There are the tourist 'migrants' of two to three weeks, people from the western (German, Scandinavian, etc.) and eastern (Sinai is still a favourite holiday resort for Israelis) urban centres, but there are also Egyptian upper-class urbanites all clothed in the latest leisure fashion and engaged in showing off their easy-going, affluent life-style. And there is the army of polite waiters dressed in black suits whose cosmopolitan, well-trained urbane conduct towards their customers gives the impression that they are playing the same game, that they are perfectly familiar with the rules of modern life even though in reality they are anxiously hiding their often rural origins. In the background of this stage where modernity is being performed, the stagehands responsible for the material setting of the scene are discernible. Usually segregated by a fence surrounding the construction site, but visible to any attentive tourist, are the peasant construction workers who erect the hotels and recreational facilities. Cooking their meals in front of their tents in the evenings and listening to the songs of Um Kulthum from the transistor radio they try to get a glimpse of this modern world passing by on the other side of the fence in the form of young and beautiful girls and boys, dressed in fashionable bathing suits, colourful bermudas or strapless sundresses. For tourists as well as for the peasants this is perhaps the first encounter with reality rather than with its representation. While Egyptian peasants have certainly acquired a clear notion of the westerner as mediated through foreign and Egyptian TV soap operas and advertising, the western tourist, if to a lesser degree, has for his part been presented with the images of the 'Egyptian peasant' through the tourist industry's folklorization of indigenous culture.[10]

In order to present a comprehensive account of different forms of migration in the Egyptian context I briefly want to mention rural-to-urban migration[11] and the overall trend towards urbanization in Egypt. As in most developing countries, this form of migration is of such pre-eminent importance that Ibrahim (1987: 217) has called it 'one of Egypt's many silent revolutions in this century'. Ibrahim shows that Cairo's average annual rates of in-migration have been ranging between 2 and 2.8 per cent. However, it has also been estimated that Cairo's growth has come to a standstill; indeed a slight decline may well have occurred in recent years. Migration to the oil-producing societies is one reason for diverting migration streams, income prospects definitely being better. Furthermore, well-established migration channels leading directly from rural areas to the foreign working sites made an intermittent phase of migration to an Egyptian city obsolete. Cairo also lost its attractiveness because of often insurmountable housing and transport problems.

THE FLEXIBILITY OF INTERNATIONAL FLOWS OF MIGRATORY LABOUR

I have previously rejected as ahistorical conceptualizations of migration focusing on spatial mobility between two distinct poles. I have likewise questioned the viability of concepts focusing on a distinct Arab migratory system, thereby implicitly making language the crucial factor delineating the territorial boundaries of a migratory system. Interdependencies are far more widespread and boundaries are not as clear-cut, impermeable or unshakeable as these conceptions suggest. Rather, flexibility is a vital constituent element whereby the landscape of migration is in a constant process of reorganization, of contraction or expansion, thus corresponding to ever-shifting politico–economic conditions, needs, opportunities and interests. If one is to speak of a system one has not only to include the Arab-speaking countries but at least also the countries of the northern Mediterranean shore, South Asia, South-East Asia and the Far East.

Academics have only recently become aware of the numerical importance of the migration of Asians to the Middle East. But migration to the oil-producing countries in the Gulf region and the Arabian peninsula has at no times been an exclusively, or even a predominantly, 'Arab affair'. The facts point very much to the contrary: right from the very beginnings of oil extraction in the 1930s Asians were involved. What is more, until 1950 predominantly Indians and Iranians rather than Arabs were employed by the oil companies (Seccombe 1988). It was therefore only in a later phase that the importance of Arab migrants in the region grew (Labib 1991: 138). Arab labour, especially highly and medium-trained experts, became of increasing interest only in the process of nation-building after independence of what had been formerly protectorates. In this process, Palestinians were of special importance, having turned into refugees in the course of the establishment of the state of Israel. Through their involvement in the former British mandatory system they formed a potential of labour power composed of high skills and administrative training which proved to be of great value for the young administrations of the Gulf states.

The oil boom of 1973, then, led to dramatically rising numbers of Arab migrants. Yet what went nearly unnoticed is that these developments likewise resulted in a remarkable influx of Asians. Here are some figures. In 1975 the percentage of Asians among the expatriate work-force of the oil-producing Gulf states is estimated at around 20 per cent (Labib 1991: 141, Seccombe 1988: 204) and around 30 per cent in 1980 (Choucri 1986, Seccombe 1988). It is also interesting to note that the percentage of Asians grew much faster than that of Arab labourers. According to Labib (1991: 141), Asian immigration in the Gulf states between 1975 and 1980 grew by 456 per cent, while that of the Arabs increased by 'only' 187 per cent in the same period. According to one estimation, a stock of about 2.5 million Asians worked in the Middle East in 1981 (Arnold and Shah 1986: 3). Initially, Asian labourers originated predominantly from South Asia, especially from Pakistan and India, but by the late 1970s growing numbers of workers from South-East and East Asia arrived. These come predominantly from Korea, the Phillipines and Thailand, but also from China and Indonesia.

This trend of the increasing importance of Asian labourers to the detriment of Arabs seems to be even more pronounced since the Gulf crisis of 1990–1. Whenever possible, Asians are preferred to Arab workers. I noticed a perfect illustration of this during a visit to a small mountain village beautifully overlooking the town of Taʿiz in North Yemen. Taʿiz attracts a considerable number of the tourists visiting Yemen, and as accommodation facilities are not yet abundant, several new hotels were under construction in early 1992. When talking to some villagers they pointed to a huge luxury hotel under construction at the edge of a rock from where one has a fascinating view over the entire town laying beyond. This terrain until then had been occupied by small shops and houses belonging to the villagers, I was told. A large signpost indicated that this was a 'development project' funded by the 'Abu Dhabi Fund for Arabic Economic Development'. Most striking to me were the Indians I could see on the site – approximately 200 workers I was told. The Indian contracting firm obviously had brought its entire staff from the Indian subcontinent. Villagers related to me that these labourers were working in three shifts round the clock and that no local workers had been accepted with the exception of some skilled manual labourers. They were well aware of the fact that the Indians were earning less than what a Yemeni would expect to get.

To understand the full implications of this account one has to remember that Saudi Arabia expelled an estimated 1 million Yemeni migrants during the Gulf crisis. This has placed the Yemeni economy, already suffering from various serious insufficiencies and among the poorest in the world, under great strain. Though no evaluation is available at present, many remigrants seem to have been reintegrated in agricultural areas. But it is no secret that a large number of remigrants, especially those who have spent long periods abroad, could not be reabsorbed and many were not able to find a new job in their home country. The real dimension of this problem can be figured out when considering that camps have been erected in the region of Hudaida, the principal industrial town of the North, to accommodate these persons and their families. With the help of international agencies the government has organized vocational training courses, medical services, etc. for these groups.

The case of the Indians in Taʿiz is significant not just as an illustration of the employment trend concerning Arab and Asian migratory labour. What is equally important is that here we find clear counter-evidence to the postulate that a labour force is cheapest where available in abundance; that migration occurs to labour-power-poor, capital-rich economies; and that eventually migration leads to an equilibrium. Nothing of the sort has happened here. Another assumption imposes itself on the spectator: the migratory system, beyond territorial boundaries, has today obviously gained such a high degree of flexibility and sophistication that not only are different international migrant groups interchangeable but also, where convenient, migratory labour ousts local labour.

Looking for reasons other than costs for the Gulf entrepreneurs' increasing employment of Asians to the detriment of Arabs, we can point to maximization of flexibility in the adjustment to the needs of the employer as one of the most important rationales. To return to the case cited above: I visited the place during Ramadhan when Yemenites, if compared to other Muslim societies, show a very low profile during daytime. Not so the Indians who were busy doing their jobs. Indians do not hamper a round-the-clock working process through fasting in Ramadhan; they do not bring production to a standstill in the afternoons through extensive qat[12] sessions; there are no families waiting for them at home in the evenings. What I have mentioned here from an implicitly structural and functional point of view is also described by Labib, though in a more essentialist way: 'Asians are considered as accepting any employment, for any price and under any conditions. For the employers it is a docile and acharnée au travail workforce' (Labib 1991: 145).

Asian migratory labour is often far more easily manageable than Arab labour. One reason is that migration of Asians does not take place individually as it does not occur outside officially established and institutionalized channels. Migration of Asians usually takes the form of contract labour where contracting firms are responsible for the entire process of labour recruitment and the organization of the work-camp. Thus entrepreneurs increasingly opt for contract labour, where periods of stay abroad are limited to the needs arising and the responsibility to carry through repatriation after the termination of the project is vested in the foreign contracting partner. These firms look for the cheapest labour on the (world) market; they organize their transport; they provide for their housing, nutrition, medical treatment, and recreation. They thus prevent community formation, perhaps the most unwanted side effect of migration. They seem to reach this aim more effectively than any immigration legislation could do. Very interesting in this context is an article by Park (1991) dealing with the recruiting practices of Japanese and Korean contracting agencies. The system is probably most highly developed in the Republic of Korea where private contracting companies provide complete 'project packages' for Gulf-based enterprises. He describes how these firms increasingly shifted from employing Korean labour power to the recruitment in other Asian countries, especially of Indians, Thais, Bengalis, Pakistanis and labourers from Sri Lanka. Park shows how, for example, the costs of an Indian labour force amount to a mere 60 per cent of that of Koreans. This trend mirrors the highly elaborate recruiting practices of such private firms which today optimize labour costs (including wages, air-fares, costs for nutrition and medical treatment) and demands of skill. What is more, Park states that these agencies now have their own regional branches in Cairo, Kuwait and Athens to observe markets on the spot.

Today capital has a global labour market at its disposal and mediating agencies have long since realized the opportunities to make profit by engaging in this 'migrant-business' and by increasingly rationalizing it. Arab migratory labour still predominantly takes place individually and is not regulated and controlled by such private firms. And Arab labour power generally seems to be more expensive. These facts serve as explanations as to why Asian labour is often preferred to Arab labour. How, then, do Egyptian migrants fit into this system? Quite simply, of all Arab migrants it is the Egyptians who seem to share most similarities with migrant labour from Asian countries. This is clear from an article by Shah and Al-Qudsi (1989) which investigates the changes in the composition of Kuwait's migrant community by comparing the national censuses of 1975, 1977/1979 and 1983. The case of Kuwait mirrors the dramatically rising weight of Asian labour within the composition of foreign labour: while between 1965 and 1975 it amounted to 29–30 per cent, in 1985, for the first time, Asians became a majority with 52 per cent. Egyptians also provided an important potential labour force. In fact Egyptian males, with 24.4 per cent, represent the second (with Kuwaiti males[13] – 24.6 per cent – occupying the first place) largest national labour force group according to a survey of 1983. Indian, Pakistani and Bangladeshi males together made up 21.6 per cent while the second largest Arab male group, Palestinians and Jordanians, constituted 13.1 per cent in the 1983 survey (Shah and Al-Qudsi 1989: 34). While the percentage of male Egyptians in Kuwait's labour force has increased during the decade under review their status within the work-force has deteriorated. Weekly working time increased by 7.4 hours, periods of stay decreased, dependency rates (i.e., the number of accompanying family members) sharply declined, salaries also decreased as did their average level of education. In accordance with this is the decline of Egyptians working in professional occupations and the sharp rise of those working in production-related or blue-collar jobs in general.

What is interesting with regard to my argument is that this gradual degrading of the Egyptian workers in Kuwait brings them ever nearer to their colleagues from India, Pakistan or Bangladesh and distances them from other Arab migrant groups. According to the data of the 1983 survey one can conclude that, *cum grano salis*, Egyptians and their colleagues from South Asia form the bottom of the occupational hierarchy of non-Kuwaitis. They concentrate in the lowest paying jobs, they consequently earn least, they are the least educated, they work for the longest hours, their duration of stay is shortest, and their dependency ratio lowest. For both groups a marked trend away from qualified and even highly qualified jobs to manual labour and blue-collar jobs is noticeable. With respect to flexibility and wages, then, Egyptian and Asian labour forces are preferred by employers. Yet, Egyptians, while sharing many characteristics with Asian workers, have one considerable advantage: they speak Arabic. As short periods of stay and low dependency rates may be interpreted as indicators of flexibility, it is understandable that Egyptians are preferred if one is looking for an Arabic-speaking labour force – which is in addition to their cheapness.[14] While no data is available, much points to the tendency that after the Gulf war the politically undesirable Palestinian and Jordanian labour force might be replaced by Egyptians and Asians. Thus Fergany spoke of '[T]he near liquidation of the 400,000 strong Palestinian community prior to the invasion to about 50,000 only at the beginning of this year' (Fergany 1992: 17). He further notes that 'the tendency seems to be to reinforce the previous relative preference for non-Arab labor. For example, one year after the liberation of Kuwait, the number of Egyptian workers there had not even reached the pre-invasion level' (Fergany 1992: 17).

Flexibility, as would be expected, is organized according to the needs and interests of the employers, not to those of the migrants. Resilience in the deployment of the migratory labour force is part of the intended optimization of migration. Any flexibility the migrants themselves could capitalize on in the pursuit of their goals is carefully sought out and eliminated. Thus it is no paradox at all that increasing market integration and the enforcing of national boundaries go together. It is absolutely no paradox at all that the free flow of labour is increasingly rationalized by channelling it through contracting firms while simultaneously the nation-state's political, and especially its legislative and executive, institutions constantly introduce new measures to impede the individuals' unhampered crossing of boundaries.

CONCLUSIONS

Summarizing my main theoretical arguments concerning macro-structural implications of migration I have stressed historicity. Spatial mobility over considerably large distances has probably always been a factor in human society. Yet it can doubtless be said that migration has only come to be of strategic importance within the process of world market integration. One constitutive factor of global market integration is the coming into being of a market of labour power and here, migratory labour force is one distinctive element articulating with other forms of labour. If we consider rural peripheral areas we can thus argue that within the process of market integration the peasants' labour power has become a commodity for external markets just like their cash-crop products. What is important is that their labour power is deployed according to exterior, ultimately world market, interests, and here the fact whether they work on the plantation of a landlord of their village community, on the tourist resort construction site in their own country or in the oil economy of the Gulf states is more or less an issue of varying distances between the peasant household, the place of (re)production of the labour force and its place of employment. This is all the more the case as at the end of the twentieth century we are witnessing a dramatic development of mass transportation through the expansion of air traffic and of other means of communication in general.

I have further stressed flexibility as one crucial factor of this globally organized system of labour supply. I have pointed out those mechanisms by which maximum resilience in an optimal adjustment of labour power to the needs and interests of the employer is achieved and maintained. The 'migratory system' is not static and it is therefore also questionable to concentrate on regional or national boundaries. There is no distinct 'Arab migratory system', as migrants of different nations are recruited, set free and interchanged with increasing rapidity. The importance of Asian labour power in the Middle East and the Gulf region is an important instance in this respect. With the accelerated sophistication of the global 'migrant business' national boundaries would lose importance were it not for their functions of warding off 'undesirable elements' and acting as a means of regulation.

It is noteworthy in this context that migratory labour power from Egypt displays many characteristics which make it often the most preferred one among Arab migrants in general. Egyptian migrants have more desirable characteristics in common with Asian migrants than with any Arab migrant group.

The system seems to have acquired such a high degree of flexibility today that virtually any desired combination of the characteristics of a labour force is available: labour power from a wide range of countries, with a wide range of qualifications and skills; working for different periods of time; female or male; different religious affiliations; a youthful work-force; varying requirements of nutrition; contract labour or not; labour force for the 'formal' or 'informal' sector – all are available on the market.

Looking for explanations as to the attractiveness of the labour force from Egyptian peasant households on global markets I have not considered an oversupply as an explanation. Instead I put the focus on its very conditions of production, that is, on the subsistence-production-oriented peasant household. This unit is the place of production of the labour force – some of it deployed on local markets, some turned into migratory labour power and sent to urban or foreign markets – and agricultural produce which is either delivered to the market or kept for self-consumption. It is at this level, then, that we meet another important aspect of flexibility as these peasant households, from the time of the *'izba* in the last century until today, are to a large degree autonomous in the organization of subsistence production and the internal allocation of the household's labour force.

3

TRACKING DOWN GLOBAL FLOWS IN THE THIRD WORLD VILLAGE

I have argued that the world system can be understood as being constituted of global flows of various types, paralleling each other, intersecting, streaming in opposite directions, growing in strength, drying up at times. For the most part such streams are associated to commodities – and here labour power is only one form – or capital. Yet, cultural aspects are surely inherent in most of these flows. Consumer durables, for example, conjure up images of a modern life style.[1] This is certainly the case in the metropolitan context, but comparable matters are also observable in the Third World village. But this is of course only one perspective from which to look at cultural flows. Another would be for the men and women involved in migration to make sense of migration and of their new environment, and that they attribute meaning to it. Without doubt people in the host society are also engaged in similar undertakings. I want to deal in this chapter with the cultural dimensions of global flows, their interlinkages with migration, and their appearance on the Third World village stage. I shall do so first from a theoretical perspective.

But to me, investigating into the facets of cultural global flows on the Third World village stage also brings questions of methodology onto the agenda. One might perhaps also deal with cultural flows on the macro level and count, for example, the number of films, soap operas, advertising spots or video clips produced in a cultural centre like Bombay or Los Angeles and sold at some other place. Or, to remain closer to the empirical setting of this study, one might count electrical consumer articles purchased from remittances and thus measure the degree of consumerism or westernization achieved by labour migration. However, to me, it is more interesting to have a closer look at the places of production and consumption. If we are dealing, then, with issues of the production of meaning we will need to engage in some sort of hermeneutic project. No longer can we deal with statistics provided by some official institution, computerize them and make sense of them with the help of SPSS, doing this at some place very distant from where the things we talk about are really happening. Instead, we should go to the people we want to learn something about and forget, for a while, about our own making sense of the facts (and about what we suppose to be the facts). Instead, we should seek to comprehend the understanding of these people. The second part of this chapter will therefore depict my own first endeavour in this respect.

TOWARDS AN ECONOMY OF GLOBAL CULTURAL FLOWS: PLACES OF PRODUCTION, CONSUMPTION AND VALUATION

Given the general emphasis of the paradigms of modernization and dependency theories on economic and structural aspects it is perhaps no surprise that culture is not a major point of interest in most migration studies. How people make sense of migration, how out-migration may be one instance of their integration into a wider world, is generally not picked out as a central theme. If it is, as for example in the discussions about the spread of mass consumer durables through the spending of remittances, it is seen in very much the same way as in both the modernization and the dependency paradigms – namely, as the global assertion of western values. Not surprisingly, this spreading within these two frameworks is assessed in quite opposite ways.

Modernization theorists tend to view these cultural effects positively. This perception had been already apparent with regard to one of the most important modernization scholars working on the Middle East, Daniel Lerner. In his famous book about the 'passing of traditional society' (1958) he predicted a kind of triumphal march of the 'rationalist and positivist spirit' (Lerner 1958: 45) into the traditional and stagnant societies of the Middle East. Perhaps not surprisingly he attributed to mobility, especially mental, but also physical mobility, a fundamental role in the Middle Easterners' inevitable taking over of western values – and this was meant to be a synonym for modernity. Within the dependency framework the globalization of western cultural values was seen as a threat, conceptualized under the heading of 'cultural imperialism' or 'cultural dependency'. In this sense, Fanon (1967) spoke of the 'colonized personality' and an overidentification with the west. Neither do world system theoreticians make a major point of cultural aspects of the world system. And it is no secret that, for example, Hopkins and Wallerstein do no more than pay a kind of lip service to such issues when they acknowledge that 'besides the specific "economic" (division of labour) and the specific "political" (state formation) there is still a third important aspect of the modern world system – the "cultural"' (Hopkins and Wallerstein 1979: 152; author's translation). Therefore, Boyne still observed, quite recently, that Wallerstein presented cultural aspects of human life as merely 'a mechanical reflex of world-systemic economist structures' (Boyne 1990: 61).

It is only in the 1980s that culture (re-)appeared on the research agenda of social scientists as a major focus of interest in its own right. In some recent publications culture within the global system has been conceptualized as one sort of constituent flow within the world system, besides others such as capital, labour power or information. Examples are the quotations from Mitchell and Sassen in the preceding chapter (see pp. 21 and 27). Within this discourse the taken-for-granted world-wide victory of western cultural values as a kind of global 'high culture' is no longer accepted unquestioningly. The very effects of such global flows have been put on the research agenda: how can we more appropriately come to terms with the obvious phenomenon of global culture in a time where market integration does not only refer to the spreading of mass-produced commodities but also to the rapid development of mass media and transport systems? All these flows have led to a situation where participation in cultural consumption has been, so to speak, 'democratized' as it is no longer restricted to the elites. But does this forcefully lead to an 'Americanization' of life-styles? Of course, mass culture and consumerism have been intruding into the lively practices of the people living even in the most remote village of the periphery: here, Coca-Cola has been very thoroughly and successfully ousting indigenous fruit juices and beverages at any important and highly symbolic occasion.[2] But, this example immediately reveals the opposite of cultural homogenization as there is an equally observable trend in this respect for 'indigenization' (Appadural 1990), of going back to the roots, or of reconquering the realm of home subsistence production. Pure and freshly pressed fruit juices are integral parts of all breakfast buffets at top-class hotels, and they are also offered at middle- and upper-class society events. Already these examples present counter arguments against the 'homogenization' approach. Within this relatively recent discourse it seems to be agreed upon that 'there is little prospect of a unified global culture, rather there are global cultures in the plural' (Featherstone 1990: 10). And Friedman has maintained that '[E]thnic and cultural fragmentation and modernist homogenization are not two arguments, two opposing views of what is happening in the world today, but two constitutive trends of global reality' (Friedman 1990: 311).

It seems to me that the alleged cultural globalization, rather than depicting reality, is itself part of a struggle over the universal validity and acceptance of a certain kind of global 'high culture'. This battle is fought for the creating of a worldwide accepted reality that 'you can't beat the feeling' of drinking Coca-Cola – whether you live in a Third World village or in an urban slum area in the centre. But as the worldwide acceptance of this feeling is the content of an ongoing struggle rather than an empirical fact the question arises as to what happens at the starting and terminal points of the global cultural flows. The empirical fact is that Coca-Cola has been spread worldwide, but we do not know much about the 'feeling'. The question is thus similar to that raised with regard to the economic, or migratory flows. Places of production rather than circulation come into the perspective. Much seems to suggest that an establishment of the signs of modernity at the apex of some globally accepted hierarchy of values does not take place without dispute or bargaining. My point of departure here is that actors, rather than taking the superiority of modernity for granted, integrate its images into their own private or collective strategies – and these may be much more sophisticated than bluntly demonstrating higher status by exhibiting as much modern consumer durables as possible. Social actors may adopt a quite creative and pragmatical approach and explore the usefulness of such symbols for their own interests. In this context it is only one possibility that the use value of a commodity is its sign character, its 'show-off' effect (Featherstone 1991; Stauth and Zubaida 1987). Whether one wears a traditional cotton peasant *gallabeyya* or a ready-made, urban-style, nylon dress; whether one builds a house of traditional mud bricks or white lime bricks; whether one bakes one's own peasant bread or whether one buys bread in a bakery – everything gives expression to difference and distinction. It is not only the needs to clothe, to house or to feed which are satisfied; these commodities also function as expressions of a certain self-stylization. And, as cultural flows are now global, and as individual or group interests differ, everybody is participating in this cultural battlefield over the establishment of the 'right' and valid mode of interpretation of signs – and those living in the traditional Third World village, where not much seems to have changed (as otherwise it would no longer be called a 'Third World' village) are in no way excluded.

In this context the question of the interrelation of international migration and global flows comes up. Given the framework outlined above a focus on the correlation of migration and the introduction of 'modernity' to the village seems to be too restricted and must remain beside the point. Viewed from this perspective migrants are not the predominant, let alone only, medium to introduce modern life-styles to the village. Rather, migrants and their households participate in this struggle, bringing in as an asset what they have acquired abroad in terms of knowledge and commodities. At any rate, village cultural life has not come to a standstill during their stay abroad as cultural flows have independently found their way to the community. Villagers are not dependent on the migrants as experts in modern life-styles. In a way everybody has developed into virtuosi in the reading off of the signs of the progressive and the backward, the rural and the urban. This position is therefore no longer the privilege of the local elite with its contacts to the outside world.

As has already become apparent, among the major 'weapons' on this cultural battlefield – whether between distinct groups or individuals inside the village, between rural and urban areas, between researcher and 'respondent', or between tourist and waiter – are the notions of 'modernity' and 'tradition'; more precisely of 'modern' and 'traditional' life-styles constructed as oppositional, mutually exclusive values. In this respect it is important to realize the ideology of this duality and the importance it assumes for the social actor's strategies and marking of identity. This oppositional duality was already a basic element of Lerner's argument when he spoke of the 'tensions [being] everywhere much the same – village *versus* town, land *versus* cash, illiteracy *versus* enlightenment, resignation *versus* ambition, piety *versus* excitement' (Lerner 1958: 44). This duality, then, has proved to be of considerable vitality and longevity, as is definitely also the case with international migration of either the 'labour' or 'refugee' sub-category. This is perhaps no surprise as the numerical importance of global migratory flows is dramatically on the rise. So also is the understanding that migrants come from 'traditional' regions – which are either economically, politically or culturally backward, lacking a modern industrial infrastructure or not knowing a democratic system and a civil society.[3]

This dualism of modernity and tradition, or modern and traditional life-styles, is not restricted to the scholarly or public discourses. Persons coming from the village (i.e., from the 'traditional' context *per se*), have definitely also integrated it into their repertoire. Politicians have done so too. This is readily noticed from two examples I want to cite here. Furthermore, these examples are interesting as they also point to the fact that cultural flows are not only directed towards the village but that they also emanate from the rural context, making their appearance in the city and even at the centres of political power. These instances are important for still another reason as they demonstrate the diametrically oppositional 'valuation' of – in this case – tradition: while in one case it has been positive in the second it turns into the absolute negative. Thus, Anwar as-Sadat, the late president of Egypt, began his autobiography with the following sentence: 'I, Anwar el-Sadat, a peasant born and brought up on the banks of the Nile – where man first witnessed the dawn of time . . .' (Sadat 1978: 9). Even though we might suspect that ultimately interests of political legitimation are at stake here, right from the start of Sadat's book it has been made clear that being a peasant is the central point of his identity (his book is called *In Search of Identity*). Surely, here tradition is positively valued. A sort of 'invention of tradition' (Hobsbawm and Ranger 1983) or 'imagined community' (Anderson 1983) is constructed. Precisely the opposite valuation is apparent in the following event. Riding one of the usually awfully cramped public buses in Cairo, I heard the bus driver, unnerved by the equally habitual traffic jam, burst into incessant insults of his fellow drivers, like 'go ahead, you *fallahin*!', or 'You *fallah*, don't you know how to drive in a street?' Yes, we all understood: Egyptian peasants know how to ride a donkey, but they have never learnt to behave in modern traffic systems, in asphalted streets with traffic lights and traffic regulations. Anybody who does not drive correctly must be a peasant and *mutakhallif* ('backward').

The suggestion is that in order to reach an understanding of what happens with cultural flows at their points of departure and termination, beyond the realm of circulation, scholarly activity must venture into the field of 'rich description: culture is that which needs to be described, that which cannot be anticipated on the basis of some theoretical premises' (Boyne 1990: 58). If we therefore want to understand something about cultural flows – and here international migration is taken to be another flow consisting not only of human and economic, but also of cultural aspects – and if we want to get an idea of how people in the 'Third World village' make sense of them, or not, we have to engage in a hermeneutic project. This inevitably means that we must leave our own secure world of the university office and scholarly community and go to the village.

'GOING NATIVE' OR 'YOU ARE WELCOME TO OUR HOUSEHOLD AS A DAUGHTER!'

I arrived at the village on a hot summer morning, together with one of my Egyptian colleagues from the American University in Cairo who was to introduce me to the ʿOmda, the village headman. It was in his household that I was to stay during my time in the community. After an approximately two hours' ride through the Egyptian Nile Delta the university car slowly made its way through the unpaved path besides the water canal up to the ʿOmda's house, raising huge dust clouds behind it and chasing poultry and children alike out of our way. The ʿOmda's house, with its three storeys surpassing all the other village buildings except for the church's tower and the mosque's minaret, was partly hidden behind a fruit orchard. While nearly all of the other village houses were one or two storeyed, built of mud bricks and covered with thick layers of straw, this house consisted of concrete and every floor was surrounded by balconies. Walking up the path through the garden and greeted by the gabbling of the geese, we reached the front-door veranda. We were welcomed and immediately led to what I later came to realize was the most 'sacred' room of the building. Equipped with 'Louis Cairo' furniture (i.e., imitation Louis XVI style), it was usually locked by key and only opened for high-status guests from outside the village, such as government officials, army officers, and the Egyptian doctor from the remote American University in Cairo. We reached this room by traversing the large veranda equipped with many palm-wood chairs lined up against the wall. This was the place where the ʿOmda usually saw his everyday guests: villagers and 'clients' who turned up to bring a matter to his attention. As I eventually noticed, the ʿOmda's wife also often received her guests at this airy place from which one has such a wonderful view over grape vines, and lemon, guava and palm trees. The veranda is the place where she in turn welcomes her more exceptional female visitors (she never uses the sitting room except for the purpose of locking away some delicious food from the covetous fingers of her children). Her day-to-day guests, neighbours or female relatives, come to join her at the back door of the house which is an ideal meeting place overlooking the kitchen, children playing in the backyard, daughters-in-law washing the large family's laundry or engaged in the tiring job of baking bread.

We were offered a second breakfast consisting of all the good things the women of the household were proud to produce: *ʿaish*, the paper-thin, dry peasant bread; *ishta*, a sort of sour cream; *mish*, the famous peasant cheese; eggs. There I was sitting, together with my male colleague and the ʿOmda, engaging in a 'formal' conversation: my colleague introduced me and the ʿOmda in turn assured him of his taking care of me and of my being welcome as a temporary family member. Meanwhile I heard the women's and children's voices and their laughter from the adjacent entrance hall – the place where family life takes place. I soon was to join them, as my university colleague left for a neighbouring village. The ʿOmda led me to his womenfolk and disappeared – he also had done his duty. I thus found myself seated among a group of peasant women of all ages, the scene animated by small children. The room was painted in light green, decorated by the pictures of Coptic saints and patriarchs and family members as well as the ʿOmda's certificate of appointment. The most important pieces of furniture were the *canaba-s* (wooden benches covered with large flower-printed cushions), a huge and almost never-used dining table (as everybody squatted on the floor around the *tabliyye*, the low table for meals), a colour television and a fan. Women and children sat on the benches or on woven plastic mats spread on the tiled floor.

The eldest woman among them was the first I became aware of. She turned out to be the *hagge*, the ʿOmda's aged mother, sitting there seemingly unshakeable in her corpulence, her dark eyes stressed by the extensive use of *kohl*, apparently somewhat senile and not too respectfully treated by the others. She immediately took an interest in me – and I felt occupied: she started to inform me about the family history, her illnesses, her daily routine, her diet. She then got up and, leaning on me, pushed me to her personal room adjacent to the entrance hall. There she opened her wardrobe and fetched a tiny bottle of oil which she explained to me contained the oil of the Holy Virgin. She told me that she had brought it with her from a pilgrimage to Jerusalem in 1965. She moistened her fingertips with it and then drew little crosses on my forehead, my neck and my wrists. She asked me whether I, too, was Christian, and she showed me the Italian-produced pictures of the Virgin Mary, Jesus Christ and some saints decorating the walls.

When I returned to the entrance hall I was immediately seized by another woman. She was about 40 years old and a sister of the ʿOmda, called Im Ayman,[4] the 'mother of Ayman', a widow for the past ten years and living alone with her four unmarried sons in a house of her own. Hearing of my arrival, she had come to join the scene. Like the *hagge* she unambiguously had me understand that I was to follow her: she took my arm and pulled me along with her to her own house in the near neighbourhood. Arriving there she showed me her ducks and chickens in the small courtyard and then she quickly led me into her house and into a separate room which was obviously dedicated to the sole purpose of worship. The smell of incense was in the air. She distributed leaflets to me and to the children and women who had followed us. These contained invocations to Baba Krullus[5] and accounts of the miracles he had performed. We were to read these invocations in a ritual which obviously was well known to all. Then she drew my attention to a kind of permanent miracle. Im Ayman pointed at a picture of Baba Krullus hanging on the wall, to the bottom of which a cotton wool ball was attached. I was told that if I watched carefully I would see his fingers move. Another of his miracles was the oil occurring now and then on the picture as well as on the window-pane. Im Ayman told her son to take a part of the cotton, to wipe the oil off with it and to make a cross on my forehead, the way the *hagge* had done before. Later I was led back to the entrance hall. Im Ayman fetched a nightdress[6] and a scarf and made me change my clothes. Having thus made sure that I stayed and did not return to the ʿOmda's family for the time being she switched on the TV set so that I could watch an Egyptian soap opera very popular at that time. She eventually engaged in the preparation of lunch: fried potatoes and eggplants. She offered me a tea, a biscuit. She showed me how she had preserved *fsikh* (raw fish in salt) and explained how much cheaper this was than buying it from the market. Again and again she welcomed me to the village, to her home. Im Ayman and her teenage son started to bombard me with questions about my religious beliefs and practices. They related enthusiastically the miracles performed by Baba Krullus and Mari Girgis.[7] In the course of the conversation she carefully tried to get to know whether I had any influence in the recruiting of the interviewers our research team had hired for the study. She finally openly asked me why her son Ayman had not been considered.

Then, to my embarrassment, my university colleague turned up and found me sitting there, dressed in a nightgown with a head-scarf wound around my head. His turning up made me feel ashamed, as if I was engaged in a masquerade – which indeed it was. Looking through his eyes my get-up must have appeared so ridiculous. However he did not turn a hair and muttering that I evidently had already been well accepted by the villagers he quickly fled the scene and escaped to distant Cairo, abandoning me to my fate. This was obviously too much participant observation for him.

Yet at that very moment I was not quite sure whether it wasn't too much for me either. With the move from the ʿOmda's fashionable sitting-room to the interior of his house I had left my own, familiar world. I had entered a totally unknown territory and at that moment I had no idea of how to make sense of what was going on around me. I had difficulties in understanding the people's local vernacular which differs very much from Cairene Egyptian Arabic. I did not at all feel sure about the intended meaning of what they said, and, what was more, even their gestures and body language seemed incomprehensible to me. These women got close to me, they did not hesitate to tap me on the shoulder and back, to grab me by the hand. When they addressed me, their words seemed to be ironical, mocking, loud, even rough, rather than friendly or polite – the behaviour I would have expected. I felt excited and curious but there was absolutely no safe ground from which to engage them in a decoding process. The only thing I felt I could do was to follow, to obey to their rules, to let myself be carried along.

I finally got back to the ʿOmda's family and had lunch with them. Then Samia, a peasant girl of about 17 years, turned up. She had previously been chosen by the ʿOmda to accompany me on my first walks through the village. As I intended to work on international labour migration she was supposed to introduce me to the migrant families of the village. Then I was to go through that morning's experiences once more. Samia, in a very polite and respectful manner, told me that she was ready to assist me and that she would join me and help me with whatever I wanted to do. She then offered to serve me the meals in my room and she did not leave any doubt that she saw it as her duty to stay with me overnight as she did not want me to feel lonely and afraid so far away from home and my family. I did understand her arguments – the individualism and wealth of the westerner sleeping in his/her own room being unknown to her. I do not doubt that she very much preferred the warmth of her younger sisters beside her at night. But this was definitely too much for me. I was at considerable pains to insist that I had decided to have my meals with the ʿOmda's family and that I had no difficulty at all in spending the nights alone.

She eventually brought me to her father's house. He was a landless peasant, who himself had been a migrant some years ago, with a small plot of rented land. However, instead of any other family members two other young girls turned up, obviously Samia's close friends. There I found myself in another peasant family's sitting-room. Though obviously this family was rather poor this room was also a kind of 'sacred' place from which everyday life is kept outside. Its comparably very modest furnitures clearly bespoke of status differences. The walls were not whitewashed like in the ʿOmda's house but consisted of mud-plastered mud bricks. Here the floor was not tiled but of pounded mud. The room was barely furnished with wooden benches covered by cotton cushions. It was almost dark inside as the only paneless window was closed by a shutter to prevent mosquitoes, flies, and the heat from intruding. What these 'sacred places', whether of the poor or the rich, have in common is that they seal off visitors as well as their hosts from mundane life and frame them in a kind of exceptional, sterile atmosphere. Rather than appreciate the honour these girls did to me, and admittedly rather impatiently, I longed to see what was happening outside. I wanted to see where Samia got the tea she served, who had prepared it and how. And I wanted to finally meet my 'research object', the migrant family. By inviting me to the sitting-room the family had shown the utmost politeness, but I definitely felt uncomfortable – especially as the girls immediately made religion the topic of conversation. I felt virtually bombarded by the three of them inquiring into my own religious practices. With shining eyes Samia's friend related how Baba Krullus had performed a miracle when he let her pass her final school exams. They also had already asked the priest whether I was allowed to receive the Communion with them and so proposed that I visit the village church to meet the priest.

The account in my field diary for what happened next on this very first afternoon in the village runs as follows:

> Later we finally had a walk through the fields and Samia surreptitiously drew my attention to houses inhabited by Muslims who seem to be living more or less in a quarter of their own. According to the girls' descriptions Muslims appear to be stupid and limited. Samia is angry that the ʿOmda chose a veiled *mutadayyina* (pious Muslim woman) of all people as another research assistant.

There I was at last on my first night in the village: alone in the Louis Cairo bedroom of one of the ʿOmda's married sons, who had been told by his father to move out with his wife and his three small children to another far more modestly furnished room. On the balcony belonging to my room I took a long breath of the fresh night air. A peaceful scene in the dark, faintly illuminated by the moon and the stars. After so much talking silence reigned at last, only broken by the croaking of the toads, the concert of the crickets, and a diesel-fuelled water-pump in the distance. Not that I felt unhappy or wanted to run away, yet I really was exhausted – not used to non-stop communication and trying to find my way into a female world to which I was a complete stranger. I admit that I felt very uneasy in what, to me, was a rather pitiless questioning about my attitudes towards religion. This bothered me not only because I am not a religious person but also because I hesitated to admit my atheism. Their interviewing of me really got under my skin. I was about to experience social death.

It took me some time to develop an idea about what had happened on this very first day in the village. What I immediately had to learn was that once one has opted for a qualitative approach there is no way around participation. I was literally forced by anybody I came into contact with to act, to react and to inter-act. I had to build up a position of my own in order to look after my own interests and objectives. Right from the very first moments my research 'objects' proved to be 'subjects' who were anything but passive and reactive in the 'research process'. Rather, it was me – on this first day and for some time to come – who had to adjust and to get to know the rules and to conform to them. I had come to the village with a clear, if abstract and scientific preunderstanding of what international labour migration is. Not that all my theoretical concepts were to turn out untenable or wrong, but they were of no immediate usefulness in my interactions with the villagers. What I had unconsciously expected was that with this kind of blueprint the traces of migration would be easily decodable once I entered the village – very much as were the famous modern-style houses in the Egyptian countryside known by everybody to be migrant homes. And, I had somehow naively taken it for granted that my role as the one who asks would be readily accepted and that my questions would be willingly answered – moreover, in a way making sense to me. Nothing of that kind had happened on this first day. Rather, I found myself struggling for survival as this world turned out to be totally strange to me with virtually nothing I could hold on to. I was completely disoriented – which was more than the feeling of holding in my hand a tiny piece of a puzzle or mosaic without knowing its place nor the dimensions of the whole – and was part of this unknown panorama. Something else confusing my perception was that although I was so fixed on my intent to unearth some underlying structure of migration and its mechanisms I seemed to be confronted with everything but that. Rather, I was involved in everyday life, its discourses reigning at a certain historical moment, in a certain village with a distinct societal make-up, in the sub-group of Christian peasant women and their perception of the world around them. And, equally difficult to accept, I felt that I was questioned and observed to a much larger extent than were these women themselves. I do not mean that at any time I was afraid or wanted to return to safer ground or that people did not accept me or did not like me. Very much the contrary was true. Yet I slowly had to learn to make sense of, and to decode, the symbol systems of this world I had chosen to join rather than looking for the representations of my theoretical concepts in the village community.

As I was not a visitor for a day or two, and it was known that I would stay and live with the women for some time, it was clear that my position within the community had to be defined and negotiated. Very clearly that was not my 'private' affair but that of all those I was to establish relationships with. Therefore, my own status *vis-à-vis* the women I interacted with had to be defined. But there was another important issue at stake. This was the possibility of their endeavouring to make me one of them in order to use me for whatever aims. On that very first day both my status *vis-à-vis* these

Christian women and the use to which our relationship might be put were still open and negotiable. Thus, rather than finding the traces of migration, I immediately realized that I was being confronted with the ramifications of my own identity as a migrant, an outsider.

I myself was part of the research process. The question as to whether I should participate or not was not an issue over which I could have decided alone as the women had already answered this in accordance with their own understanding. When I came to the village I immediately became part of an ongoing struggle over legitimacy and the hierarchy of symbols: not only my own legitimacy, and that of these Christian women in relation to me, the western, 'modern' woman, but also that of the women in relationship to the Muslim community of the village – all were at stake, tested out and negotiated on this very first day. One of the most important symbols in these interactions was my own identity as the migrant, further characterized as the western, 'modern' woman. I was a woman from West Germany, known as one of the major paradises of modernity where everything was 'clean', 'developed' and 'advanced' – where virtually any villager wanted to go. I also had a university degree. I wore modern clothes, I had short hair, and I did not yet have children. I was free to decide about where I wished to live. But what in other contexts may have been signs of 'progressiveness' or female 'liberatedness' were not admired by these women; rather the contrary seemed to be the case. I was the one not conforming to the rules and values. I had come along to do research work – but was I married or not or where were my father and mother, my family? Did I not have any children although I was already in my early thirties? Why did I, a woman from the rich west, not even wear any kind of gold or jewellery? What, to study migration? What for? Does that feed anybody? Yes, definitely I was a symbol of modernity – but I really did not feel that this would at all enhance my status *vis-à-vis* these women. They felt pity for me.[8] Here I was in another territory with other rules and the only thing I could bring forward to justify myself was that things were not the same where I came from and that they had to accept me as I was.

Entering village society I at once found myself on an ideological battleground characterized by the duality of modernity and tradition. It took me some time to understand that these Christian women instantly engaged in a kind of multidimensional strategy: this was the bargaining for status between themselves and me. Here, it was exactly these signs of modernity they called into question (i.e., my travelling alone without a husband or family for the sake of a job, my childlessness, my being short-haired, my not wearing gold), thus underlining their own moral superiority as the ones conforming to the superior community's traditional moral code. In the light of the villagers' negative evaluation of my attributes of modernity they made a positive assessment of their own role as traditional peasants in a Third-World, subsistence-producing village. And here, village tradition is positive *per se*, justified by the fact that 'things have always been that way', the 'this is our village custom'. I was the one not conforming to the village traditions' moral code.

On the other hand, the setting, carefully arranged by the hosts, indicated their interest in being considered equal, and knowing about my modern life-style: the dirt of the village had to be swept away. Villagers, in their role as hosts, created an exceptional atmosphere, as I have already shown earlier when describing rooms where 'high-status' or rare guests are welcomed. Yet the creation of this atmosphere is not restricted to the choice and decoration of location. A specific set of modern mass-produced consumer durables is likewise important. Rather than the usual tea, soft drinks – especially Coca-Cola and Fanta – are offered, which are rapidly fetched by one of the household's children from the shop around the corner. Stauth (1990c: 28) reports how the electrical fan is positioned near the guest and often the TV is switched on. The endeavour to create purity is another notion I observed as belonging to the setting: the villagers attempted to eliminate anything associated with the dirtiness of village life. This ranges from large flower-print cotton covers for the precious urban-style furniture to the excessive use of *flit* or any kind of spray insecticide. I very often found myself sitting down in a peasant home while *flit* was fetched and sprayed around me, befogging me in fragrant poison clouds to ward off mosquitoes, flies and, most embarrassing to the hosts, the fleas. As I struggled to breathe the visions of frequent TV advertising spots came to my mind: easy-going, young, urban middle-class families; innocently sleeping young children in white cushions; bright colours and vibrant music – all associated with this kind of insecticide. The same advertising could just as easily have been used to promote a kind of spray deodorant.[9] This was how life in my habitual surroundings was presented to the people of the village, and by using such appliances and by creating a 'clean' space the distance between us somehow was reduced.

The struggle over symbolic hierarchies, however, was not restricted to the relationship between these women and myself. At the same time I also constituted a very welcome asset in their struggle with their Muslim compatriots. It became increasingly clear to me that this was one key to understanding the women's insistence on discussing the subject of religion that had so much irritated me in so many encounters. For sure, religion definitely plays an important role in the structuring of the women's everyday lives; more than that, it is usually associated with life's more joyous and exceptional events and also with leisure time. Excursions to monasteries and to the graves of holy persons, to the annual *mulid* (the commemoration of the villagers' most important miracle-working saint), and the ritual passages of baptism and especially marriage (but of course also death), are all celebrated with extensive festivities. And it is impossible to deny the dimension of faith in these women's religious practices. On the other hand, what I experienced on my first day in the village indicated that religious behaviour was an important aspect of an ongoing symbolic struggle between the Muslim and the Christian community of the village. Here, in the wider context of the village society, my otherness was not contested but instrumentalized as a most welcome asset: I here appeared as a representation of modernity and I was Christian. Knowing me to be on their side instead of that of the Muslim families of the village became a vital issue from the very first moment. Suddenly the women's strange and puzzling insistence on our common Christian identity made sense to me, and as at that time I was not yet identified with either of the two sides the endeavour was all the more promising. In confrontation with the Muslim community, 'tradition' was no longer taken to be a positive value. Here, it again meant 'dirt'. The Muslims of the village were relegated to the sphere of tradition, this time connoting poverty, dirt, landlessness, illiteracy: the peasantry. These Christian women used their association with me to reinforce their own modernity. This also clearly conveyed a double message. To me, the western woman, they said: 'Look at the traditional Muslim – we are not like them!', while they seemed to tell the Muslims: 'She, the modern woman is with us, not with you, as we are more similar to each other than you are!'

Here is another example. I suddenly realized the fact that Samia, the peasant girl who introduced me to the village, always, when with me, was dressed in a shining, ready-made, urban style *gallabeyya* of synthetic fibre. Moreover, when I visited her at home

I always found her dressed in the 'traditional' *gallabeyya* young women of the village usually wear. I became aware of this fact because Samia herself one day told me that she deliberately wore this dress when with me. While puzzled at first I suddenly understood that by dressing this way she somehow symbolically reduced the distance between the two of us, which was all the more important as a visual aspect when we were visiting the Muslim households of the village together (which she was never very enthusiastic about). It is perhaps interesting to note that I often wore one of these urban-style *gallabeyya-s*, which happened to be a dress brought back by a migrant from Saudi Arabia and was easily recognizable as such to the women. I again was the one having difficulties in decoding this message. At first I had put it on simply out of a lack of what I thought to be adequate dress and because in my endeavour to 'go native' I had favoured it rather than my blue jeans. It was only later that I noticed that a number of women admired this dress, its significance to them being that it was urban-style, Saudi-Arabian, and made of a synthetic fibre instead of the 'traditional' Egyptian cotton – this was modernity![10] If there is any prouct symbolizing the Egyptian peasant and his/her working with the soil, then it is cotton!

A great deal of flexibility reigns in this kind of ideological battleground in a village where opponents and coalitions change incessantly, sometimes imperceptibly, and where also the intended meaning of messages is kept vague. In this sense Featherstone has argued that:

> These images cannot be regarded as amounting to a coherent dominant ideology as images are continually re-processed and the meaning of goods and experiences re-defined. Everything becomes exchangeable with everything else and there is no apparent limit to the transferability of carefully delineated and once sealed apart meanings.
>
> (Featherstone 1987: 20)

Villagers send silent messages which are nonetheless easily understood by those concerned. It is with this virtuoso playing with different meanings of modernity and tradition that we must become aware.

4

KAFR AL-ʿISHRA

An agrarian village in the 1980s

I have stressed in the last chapter the historical continuity of a system within which peasant households organize subsistence production autonomously to a considerable degree while at the same time producing cash crops and labour power for external, increasingly globally organized markets. My focus has been particularly on the attractiveness of this organization of production on a macro-structural level. In this chapter I want to change the focus to the micro-level and deal with the present day ramifications of this system in the village of Kafr al-ʿIshra. Stress will be put on the economic activities performed by the villagers. Inherent to an economic system are relations of power and their representation which I shall deal with by showing how the built – and unbuilt, for that matter – environment is marked by signs of power and powerlessness. This also implies a struggle over the occupation of space and time. I shall show how villagers engage in self-styling and the 'construction of the other'.

GLIMPSES OF THE PAST

Kafr al-ʿIshra is not an ancient, Pharaonic village. Its origins go back to the time of accelerated market integration. Though I could not collect enough material from the narratives of the villagers to compose its concise history I think it worth while to give an account of its fragments here. The name of the village is interesting in this respect. *Kafr*, etymologically an Assyrian word (Ramzi 1954: 5), designates residential areas which have been separated from the original village. Its history dates back to the middle of the last century when in a long process of legal establishment of private landownership large estates were founded, especially in

Lower Egypt. Boinet (1899) mentions the village in his *Dictionaire Géographique* as belonging to the district of Sinbillawein and indicates the number of its inhabitants as 349 at the turn of the century. Village history of Kafr al-ʿIshra, as narrated in the community, says that it was founded by two Christian brothers from Mansura. They were fairly-well-off landowners, each of them owning about 120 *faddan.* In order to better monitor agricultural production they established residences in their fields and thus the nucleus of Kafr al-ʿIshra was founded. Successively, landless Muslim (as well as Christian) cultivators followed them and built their huts in the vicinity of the landowners. Black families living in contemporary Kafr al-ʿIshra point at another phenomenon linked to the expanding market production at that time: agricultural slavery. George's wife still remembers Michael, their *Habashi.* He was the household's slave during her childhood and worked for her family for 25 years until he died. His job had been to monitor the work of the cultivators in the fields. She narrated how much they had loved him and that he never wanted to leave them.

While one of the two brothers eventually sold his land and moved to the city, the other kept his land. He built the first church of the village and his grave is still to be found in the churchyard. He became the first priest of the community and thus he also turned into the spiritual authority for the peasants. This office still remains in the hands of his descendants. Hence from an early period of the community onwards politico–economic power was backed by religious legitimization. The priest's family is still among the largest landowners of the village today. Later two other Christian families settled in the village, coming from Mansura, the governorate's present capital. Their descendants are still among the most wealthy landowners of the village. However, they do not live here but in Cairo, Mansura or other important urban centres. They only come to collect their revenues at each harvest. Some of them work as bankers or engineers. Also, the priest lives in nearby Sinbillawein and only comes to the village to perform his priestly services and to monitor agricultural production on his family land. Kafr al-ʿIshra might have originally been *ibʿadiyya land* (from: 'to be remote') which denotes a category of land hitherto uncultivated. Muhammad Ali (1805–49), from 1820 onwards, granted it to high officials and also to village notables. These donations were tax exempted under the provision that the new holders cultivated it – usually with cotton (Baer 1962: 16). In 1842 almost complete rights of ownership, including the right to sell and transfer the property, were granted to the proprietors[1] and it must have been in these days that Kafr al-ʿIshra was registered under its name as independent from the original village (Ramzi 1954: 8). Estates established by the landlord at some distance from the original village on *ibʿadiyya* land were mostly operated in the *ʿizba* system. However, Kafr al-ʿIshra was not an *ʿizba*. Here, forms of share-cropping and cash–rental agreements must have prevailed.

Plate 4.1 Kafr al-ʿIshra. In the lower part of the picture three large pots can be seen. These are used as granaries. The two tall towers in the upper left half of the picture are dovecotes. The large light-coloured pile to the mid-left consists of hackled wheat straw which is used in the production of mud bricks and as fodder.

One still finds many traces of Kafr al-ʿIshra's past in its contemporary sociopolitical composition. The founding of the village occurred because a rich urban Christian family managed to become the local beneficiaries of the mechanisms of market integration, just as we meet the small cultivators as its losers. By and large these Christian families have managed to retain their position at the apex of the community's societal hierarchy – though there are, of course, also poor Christian households. And, the majority of Kafr al-

ʿIshra's landless or near landless families are still Muslim. Kafr al-ʿIshra did not have very large landholdings, and acreages of the rich families were thus below the margins of Nasserist stipulations; furthermore, in principle landless peasants did not profit from *islah*, the land reforms. These surely are reasons which may explain why the community, compared to others where land was redistributed, is relatively poor even today. Muslims have managed to buy small plots of land only in very recent years. Thus, the largest Muslim landholdings do not consist of more than about 5 *faddan*.

THE VILLAGERS' GRASP OF REGIONAL GEOGRAPHY

Kafr al-ʿIshra is situated at the south-eastern fringe of the Delta province of Daqahliyya. It is located on the main, asphalted road which links Mit Ghamr and Sinbillawein, the district town, to which the village administratively belongs. As hourly buses connecting Cairo to Sinbillawein pass on this road villagers can easily reach Cairo. The capital is at a distance of approximately 100 kilometres. It usually takes some two hours to get there and those who are looking for a more rapid way of transportation take a communal taxi. These taxis leave their departure point in Sinbillawein as soon as seven passengers have come together, which usually does not take a long time. Largely due to the notorious audacity of their drivers these cars indeed afford a shorter journey but numerous demolished taxis besides the highways have always led me to abstain from this sort of transport. Fatal accidents are frequent and I know several families in the village who have been bereaved due to unscrupulous drivers.

Cairo is a very important centre in the minds of the villagers. It is important because of its central administrative units, its numerous markets, its universities and especially its medical facilities, it is the living place of family members, neighbours and absentees. Cairo is also the centre of TV and film production. What is probably more relevant than production in this respect is that in these films Cairo often figures as the scene of the action and villagers are thus used to the images of upper-middle-class urban life-styles.[2] Though Cairo is now frequented by a considerable number of young men who work or spend part of their military service there, to the majority a trip to the capital city is still a journey to an outside and strange world. Visits remain reserved for exceptional occasions.

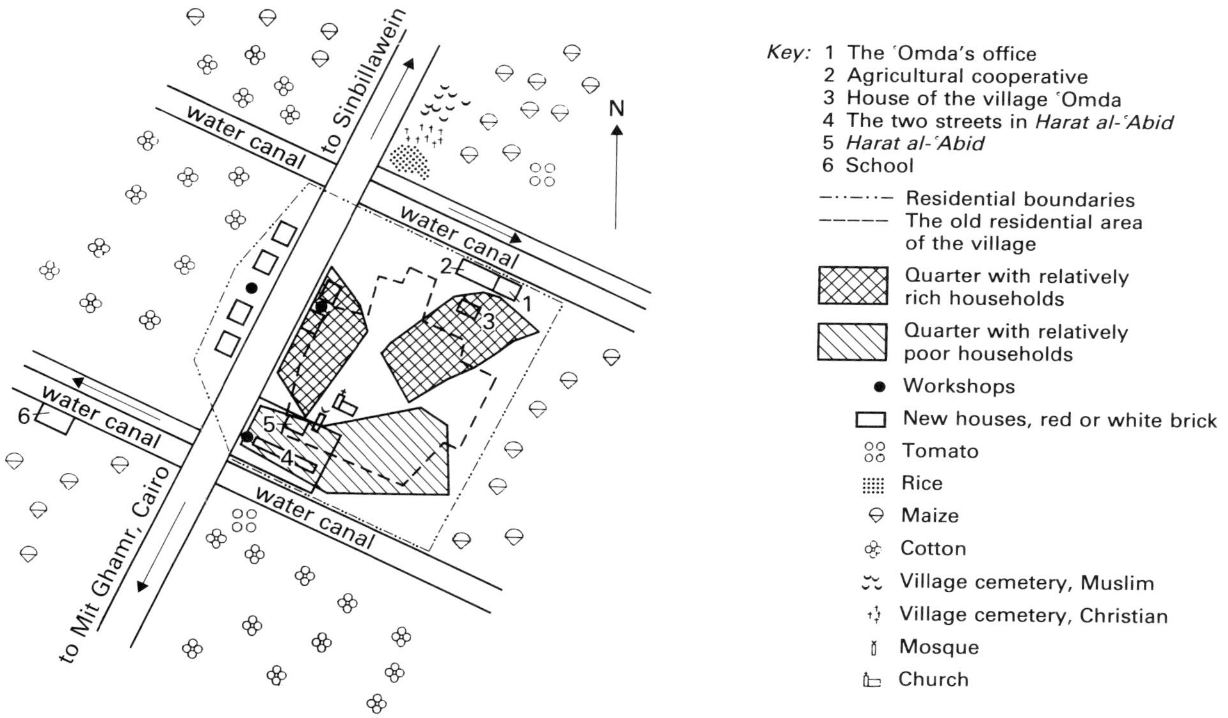

Map 4.1 The village of Kafr al-ʿIshra

This is especially the case for the village women for whom Cairo usually remains beyond reach. Most older and poor women I asked have never been as far as Cairo. Though figuring prominently in their mental maps, Cairo remains outside their practical, everyday landscapes. I recognized this when I wanted to leave the village for the first time. Announcing my intent, my guest family, feeling responsible for my well-being, got rather concerned at first. They could not imagine that I would find my way on my own without getting lost – while I, in turn, was astonished at the degree of their worries. Later I learnt about the elaborate preparations needed for a woman's journey to Cairo. Wizze, the ʿOmda'a wife, used to visit her sister-in-law in Cairo once a year. This was always a time-consuming enterprise with extensive preparations. Some days in advance of leaving she would be busy preparing huge baskets with rice, home-made bread, cheese and flour. Finally she would fetch the chicken. An acquaintance from Sinbillawein owning a taxi would be contacted, picking her up in the village early in the morning. Then the car would be loaded with all the baskets, bags and bundles, driving her right to the front door of her sister-in-law's house in al-Darb al-Ahmar in Cairo.

Even Mansura, the capital of Daqahliyya, about 20 kilometres north of Kafr al-ʿIshra and an important commercial and administrative centre, remains on the periphery of the villagers' world. But some have married children there. And those who don't have much confidence in the capacities of doctors in Sinbillawein will first turn to Mansura before eventually going to Cairo. It is at Sinbillawein, the district centre and market town, where the practical everyday life land maps of the villagers seem to end. Here the district's main weekly market is held, and medical services, schools, municipal institutions, the post office, telecommunications, wholesalers, jewellery shops, bakeries, boutiques selling ready-made clothes, 'photo studios', two compartment stores, tailors knowing how to sew urban-style dresses, are all to be found here. Visits to the district town are thus a frequent necessity for women. They go to the market or the commercial centre to procure all the household items they cannot find in the village shops – which is virtually everything except for tinned tomato paste or sardines, rice, potatoes or sugar. They will stop one of the growing number of privately operated 'pick ups', a sort of small truck to which wooden banks have been installed to receive the passengers. As there is a great demand for this kind of cheap transport these Chevrolets or

Bedfords are normally very packed. And it is usually on Thursday, the market day, that the black-dressed village women, with huge baskets containing vegetables, fish, living chickens and rabbits, or aluminium kitchen utensils, are all crammed together on the small and covered load area. Naturally, this is also an important place of information exchange as one is always sure to meet people from adjacent villages who know the latest news – thereby proof of another kind of communication or information network between the villages is provided. Transmission of village news is still linked to transport. A young woman conveyed to me the secret of her pre-marriage romance: as they had to keep totally secret she and her 'boyfriend' used to meet 'by chance' in pick-ups or buses, which gave them the time of the ride together to Sinbillawein or even Mansura.

Besides Sinbillawein, the villages in the vicinity of Kafr al-ʿIshra are often visited. Women frequent the markets of nearby villages, where sellers are thought to be more honest than those in Sinbillawein. Goods are also judged to be 'cleaner' and cheaper there – the reason is that the traders here are known to be peasants, not professionals. Women also go to visit their daughters married to men in these villages or to offer their condolences when someone is ill or bereaved. It is only the students of intermediate schools and institutes, those who do their military service, and government employees who regularly and daily participate in a wider world. And there are, of course, the migrants, whose world has become much wider than that of most of the rest of the villagers.

THE HABITAT

Kafr al-ʿIshra is a cluster of houses surrounded by the green or gold-brown of the *gheet*, the agricultural fields. Approaching it from some distance, the impression of a dusty, grey-brown homogeneity prevails. This image is emphasized by the agglomeration's lack of green colour which outside the residential areas is so pervasive with cotton, clay, wheat, rice fields, palm trees and eucalyptus abounding. However, inside the village hardly a handful of households have a small kitchen garden planted with an orange or palm tree, grapes and peppermint or jasmine.

Cars have a hard time finding their way through the generally narrow, unpaved streets – but up to now hardly any household owns a car. It is therefore mainly itinerant merchants and city-

Plate 4.2 Kafr al-ʿIshra: a village alley. Women usually carry heavy weights on their heads. She has not yet balanced out the bowl on her head as she still holds it with her hand. Usually women do walk 'free-handed'. The boy is not wearing the more common *gallabeyya*, but an urban-style pyjama.

dwelling members of the old landowning class who on their occasional visits to the village try to find their way by car through the narrow alleys. Car drivers avoid the labyrinth of the narrow village lanes and restrict themselves to the main asphalt road passing outside the village. Actually, the expression 'passing outside the village' is no longer true. A couple of years ago, on one side of this road which once separated the village from agricultural land, a row of two- to three-storey houses appeared. These have facilities for merchant enterprises or workshops located on the ground floor. On the fringes of the village this building boom is easily discernible. As free space is scarce within the village everywhere at its boundaries houses are built on what had once been agricultural soil. Some of these houses are still unfinished and lack a roof, others are apparently not inhabited – these may belong to migrants still working abroad to earn the necessary money to complete their new homes. The newly erected buildings give proof of the villagers' defiance of the government's regulations which strictly prohibit the building of new houses on agricultural land – given the extreme scarcity of agricultural fertile soil. The peasant homes consist of one or sometimes two floors and often house animals and human beings under the same roof. Houses are covered by thick layers of dried clover, wheat straw and maize stalks, which serve as a protection against winter rains and as a storage place for animal fodder and 'firewood' at the same time.

The great majority of Kafr al-ʿIshra's houses (83.4 per cent)[3] are built of the material which for centuries has been used for this purpose: sun-dried mud bricks. These bricks are usually produced by the villagers themselves. Chopped straw, mud and ashes from the oven are mixed and shaped into bricks, then dried in the sun. The production of such mud bricks for sale has been one of the major income opportunities of landless and poor peasant households, which, though forbidden,[4] is frequently practised, especially in agricultural slack times. Buildings of other materials, especially red (burnt mud) brick and the cheaper white lime and concrete stones in Egyptian villages, are one indicator of the quick changes and the trend towards urbanization which has been remarkable in the last two decades. And with the new houses, new styles and a new architecture were introduced. The 'urban' villa and apartment building have gradually found their way into the village – buildings of several floors, painted in light colours and with a clear separation of functional spaces which had hitherto been unknown. Kitchens and bathrooms are built and animals are separated from the immediate living space of the human inhabitants. Here the typical straw cover of the roofs is gradually vanishing and TV aerials have begun to appear. But while Kafr al-ʿIshra is thus also participating in the tremendous building boom taking place in the Egyptian countryside, here most new buildings are still built with the traditional bricks. The choice of the construction material is to a large extent determined by prices: to build an upper floor of mud bricks may be virtually free of expense if household members themselves produce the bricks and use the soil of their fields to do so. If these are purchased costs amount to about £E2,000, while a construction of red brick costs £E10,000 – an impossible sum for most of the small peasant households. While the rich build in red brick, the 'well-to-do' of the village are increasingly using lime stones, brought from Upper Egypt by a villager who owns a lorry. While the lime stones are cheaper than red bricks, they have the definite disadvantage of neither warding off the heat of the summer nor the cold of the winter. The architecture of buildings with lime stones does not differ much from the mud-brick houses. Floors often still consist of stamped mud and ceilings of wooden beams covered with straw.

The social life of the people living in these houses also conforms to a 'rural' characteristic to some extent as some 30 per cent are living in extended and joint families, where married sons live together with their wives and children and their own parents. They jointly organize the work and share in consumption of the output under the ultimate authority of the paterfamilias.[5]

Most households have been equipped with electricity since it came to the village in 1977 – save for the newly built houses on the edge of the village and those of the very poor and destitute. These new houses are not served with electricity for legal reasons: as these are illegally built on agricultural soil they cannot be connected to the communal current network and their owners usually have a hard time until they finally manage to get electricity. Electricity has illuminated the peasant homes in a twofold way: first, all the houses equipped with it have installed light bulbs; second, it has brought the TV set and radio recorder to the home. Even for the few who do not yet have electricity or a TV set, it is always possible and very common to join a neighbouring household in one of the important events: the watching of the daily *musalsil*, the Egyptian soap opera, or of an Egyptian or foreign football match. Thus, electricity has tremendously accelerated the integration of the village into the national as well as international context: in the evening everybody is able to see which institution President Mubarak has officially opened or what foreign guest he has welcomed or said goodbye to during the day, and also the American president's delegates' latest shuttle diplomacies to solve the Middle East crisis. Indeed, enlightenment in this double sense has been the dominant result of the introduction of electricity. To date, only a minority of households use electrical current for a refrigerator, an electrical fan, an iron or a washing machine. These signs of participation in modernity, though highly desired, are still restricted to the few wealthier households.

The government has built a large well to secure the supply of drinking water, but only a handful of households are linked to it. The water goes instead to a more affluent neighbouring village. Today, and this is considered to be a success in health care, most households possess their own manually operated water-pump and a few richer households have electric pumps. The poor who cannot afford their own pump use that of their neighbours. This has not always been the case as one woman remembers:

> In the old days, all the people used the water of the water canals. They washed their clothes and their dishes there, as you can still sometimes observe today, and they also drank from the canals. We used to purify the water with ground apricot seeds. Only the large landowning families had their own well when I was young.
>
> (Peasant woman, 45, illiterate)

The availability of water-pumps to nearly every household has thus had a beneficial effect for women. In Kafr al-ʿIshra women are no longer washing clothes and dishes in the canal and they no longer have to carry water over long distances. Yet the existence of a water tap linked to a communal system of drinking water to every household is still a utopian dream. Electricity, in fact, remains one of the rare services at the disposal of the villagers. Until the end of the 1980s there was no post office or telephone, no clinic or public health station, no village bank, no social unit to which the poor could address themselves, not even a village school. It is the non-existence of such communal facilities which imparts on the observer and the concerned the impression of a 'traditional', remote, village abandoned or neglected by the state. Thus the most important 'stately exports' to the village are the state-controlled TV and radio programmes, and the state employees of the agricultural cooperative. No wonder a young man from the village complains:

> Our village is the most backward one to be found all over Egypt. All the houses are of *tub nayy* [mud brick]. We do not have a school. *ʿIzba Shlube* [a neighbouring village] is much smaller, and there have been only three or four *hara* ['quarters'] with only three to four houses. But today every-

thing there is built with *tub ahmar* [red bricks] and they also have a school!

As there is no school, not even a Quran school in Kafr al-ʿIshra, pupils have to walk for about a quarter of an hour to attend primary school in one of the neighbouring villages. Any further formal education is to be found only in Sinbillawein. It is therefore not astonishing that 76.2 per cent (n = 218) of household heads in the village are illiterate or only have some very rudimentary knowledge of reading and writing. As we are here talking about the education of the household heads, literacy can be expected to be generally lower as many of these men are too old to have profited from the Nasserist reform. But four decades of primary education, on the other hand, have not led to a substantial reduction of illiteracy. Though most children go to school, many, especially girls, drop out after only two or three, or, at maximum, five years. Besides unbearably high fees for books one important reason is that girls are needed to help their mothers in the household. For pupils of both sexes it holds that they have to work as *anfar* – as agricultural daily labourers whenever a chance for additional monetary income can be seized. They also have to work on the household's fields. Parents mention another reason for not letting children go to school. They are afraid of accidents on the asphalt road the children have to take on their way to school. And, in fact, I know families who have lost a child that way. Another striking fact is that quite a large number of children, though having completed five years of primary education, are unable to write more than their names – here neglect, but also very limited financial resources of the families, is to be quoted among the reasons.[6] Thus, it is no wonder that only the rich have an interest in and the means to afford a good formal education for their children.

THE SOCIAL GEOGRAPHY OF THE VILLAGE COMMUNITY: SYMBOLIC STRUGGLES OVER THE OCCUPATION OF TIME AND SPACE

The nearer one gets to Kafr al-ʿIshra, the more the impression of its homogeneity dissolves into an heterogeneity of distinct buildings built in different materials, in various styles, of different functions and sizes. The closer the acquaintance with the village and its inhabitants becomes, the better we can detect the signs marking its social map. Strolling attentively through its residential areas and listening to the people talking about the village community we will observe new identity markers and discover an increasing number of signs speaking of the construction of difference, of boundaries and territories. Eventually, a social landscape of unequally distributed power and influence takes shapes. Therefore, constructing difference and identity and setting boundaries has much to do with space, with its creation, occupation and defence. It is thus Kafr al-ʿIshra's social landscape that will delineate in this section.

As elsewhere the amount of space a family is able to occupy symbolizes its prestige, wealth and power. In the agricultural village dominant families have the largest landholdings, the largest houses, courtyards and gardens. Here, the ʿOmda's multi-storey house, already described in Chapter 3, is a case in point. His premises are among the largest in the village, and his home is also the highest, only being exceeded by the two buildings representing religious power: the mosque's minaret and the church's tower. His house is easily discernible from some distance and even from most places within the village. From his roof one can see the whole residential area and its fields. His estate is built over a *faddan* of land – while most of the peasant households occupy approximately one *qirat*, one twenty-fourth of the ʿOmda's estate. Also, his house is one of the very few which has a relatively large garden, planted with guava, grapes and date trees – it is even large enough to allow for the growing of bushes which serve no other purpose than that of decoration. His estate also comprises separate buildings for the storage of grain, stables for poultry and cattle, the oven, a threshing floor, and a place for the agricultural machinery. The space of the wealthy is usually enclosed and segregated from that of the 'common people'. It is therefore not astonishing to find that the ʿOmda's premises are surrounded by a wall and cut off by a gate. Privacy is thus created to a degree poor families could never afford. Privacy is a luxury and expression of status at the same time. One should perhaps also mention the internal organization of space in these rich families' houses. No poor family can afford to functionally separate a kitchen, bathroom and toilet and to build a separate stable for the cattle.

On his premises, the ʿOmda has accommodated an elderly poor couple in a separate room. This act of generosity symbolised in the allocation of space is clearly another indicator of status. It acquires its specific significance as this couple counts among the *ghallaba* of the village community. *Ghallaba* – these are the most destitute and poor who do not possess any assets to engage in any sort of subsistence production – labour, land, animals. They cannot survive without the community's charity. It is through the presence of this elderly couple that the ʿOmda's qualities as a benevolent, caring patriarch and village headman find an important visual expression. This self-styling as the one caring for the needy is duplicated by another medium, that of storytelling. Such narratives provide information about how villagers make sense of the community's everyday life and its history. On the other hand, the social reality and landscape arising before the inner eye of the listener is also to some extent the product of the imaginative power of the storyteller. It is in this sense that the narratives of the ʿOmda's wife, Wizze, and one of his daughters-in-law, Batta, as persons belonging to his extended household, about the blind woman and her sick husband, reveal their meaning:

> Laila and ʿAm Ibrahim have been married for about fifiteen years. ʿAm Ibrahim has been married before and he has two children. When his first wife died he married Laila. His eldest son threw them out of their house. They are not from here, ʿAm Ibrahim is from a village in Minufiyya. Laila has a sister in the village and this is why they came here. She also bakes her bread in the oven of her sister. They are *ghallaba*, they cannot work because she is blind and he is old and ill. Laila sometimes helps me with my household chores. They live from the villagers' gifts because they are *ghallaba*. From one family they will get 2 kg of rice, from another 1 or 2 kg of wheat. They have to buy everything on the market: vegetables, fruit. . . . The ʿOmda always gives them meat on holidays. When they came to the village for the first time they looked everywhere for a room to stay in, but nobody wanted to rent them a room. Then the ʿOmda gave them one of the poultry stables. I give them everything, meals, rice, flour. Others do the same.
> (Wizze)

Batta, the ʿOmda's second daughter-in-law, also once told me the story of ʿAm Ibrahim and Laila. Batta, it should be pointed out, is the daughter of the ʿOmda's deceased predecessor.

> Laila had been married to a man from another village. When she went to live with him she eventually found out that he

> was not Christian like herself but a Muslim. She wanted to divorce but the Church did not recognize this divorce. . . . Despite the difficulty of her case she eventually got ʿAm Ibrahim, who already had two married children and whose wife had died. He married Laila and therefore his son threw him out. They therefore came to our village because Laila's sister is living here. Laila had a child by ʿAm Ibrahim, but it died. First the late ʿOmda gave them a room. However, this room was turned into a youth club. Then they were given a room by other villagers, but this room also was needed because somebody married. Then they came to the ʿOmda, who gave them room without asking for money.
>
> (Batta)

Both women underline the identity of the ʿOmda as a benevolent patriarch protecting and caring for the needy – which is the obligation of the large landowner *vis-à-vis* his cultivators according to the *taqlid al-qarya*: the village tradition and its moral universe. As the ʿOmda's affluence very much depends on the work-force and the loyalty of the peasants he must be seen to be interested in the maintenance of the community's moral universe. Both sides, in fact, depend on one another: the destitute couple relies on material inputs from the ʿOmda's household whereas the ʿOmda needs them to style himself as the caring patron. It is interesting to note that Batta, as the daughter of the former ʿOmda, also mentions that her own father gave them shelter. The constitutive element of this moral economy is the affluent subsistence-producing household which can easily locate part of its products (rice, wheat, meals) to the needy. Here the absolute opposite – the poorest household – is the one which does not dispose of any assets required for subsistence production: agricultural land, cattle, poultry, labour power. It is here that Wizze's mentioning that 'They have to buy everything on the market' conveys its meaning. The very presence of its opposite, the destitute household, on the premises of the wealthy landowning household counts among the most powerful symbolic expressions of the 'traditional' legitimacy of the ʿOmda. Yet, he has also managed to give his power a 'modern' (i.e., adminstrational) image. When the ʿOmda took over his function he had new rooms built to house the bureau of the *ʿomdiyya* and the agricultural cooperative next to his estate. Both institutions are the most obvious representatives of the state in the village. By bringing these offices (which had been hitherto near the former ʿOmda's home) closer to his house, a spatial rearrangement with the objective of power concentration and its symbolic expression has taken place. His legitimacy as the community's headman has found twofold spatial symbolic expression.

The ʿOmda is an example for those village-based families who, through the concentration of family land in their hands, have managed to become of great social power in the community, a position which is then further strengthened by taking over a political office, that of the *ʿomdiyya*. This is interesting as he does not actually belong to the old landowning aristocracy of the village. The ʿOmda's grandfather had come from Saʿid, from Upper Egypt, to this region and it is his son who had come to considerable fortunes. The latter was *nazir* (i.e., administrator) on one of the nearby very rich *ʿizba-s*. His wife, the *hagge* (see Chapter 3), narrated to me that they had all been living on the *ʿizba* until he bought land in Kafr al-ʿIshra and built his *sarail* here. This was shortly before her son, the ʿOmda married (some 25 years ago). According to stories told in the village the *nazir* was able to buy 38 *faddan*. People narrate how he used his position to divert considerable funds and assets for his own interests. There were four *ʿizba* workers exclusively occupied with looking after the *nazir*'s cattle, there were also two women working in his private household, and there were others working in his fields. These labourers were not remunerated from the household's budget but belonged to the *ʿizba*. It was the *nazir*'s son who by marrying Wizze, a sister of George (whose family had the slave), established family relationships with the village's historical elite. It is further interesting to note that the present ʿOmda managed to avoid fragmentation of land through heritage. As the only son he refused to give to his sisters what was theirs when their father died, except for two of them who stayed in the family. One married a brother of Wizze, and one of his own daughters married a son of another of his sisters. The other three sisters did not receive anything except gifts of grain at harvest time. The land under his control also comprises that of Wizze, his wife. By taking over the *ʿomdiyya* and thereby bringing political power to himself he has consolidated what his father had begun. Today his family is among the most influential and powerful in the village.

While the ʿOmda's premises definitely are the most prominent example of a village-based, middle-class landlord's occupation of space, other examples could be given. There is still one old house lived in by two females of a major absentee family. These old houses of the formerly important absentee aristocracy of the village have a charm of their own. While today in a state of dilapidation, they still convey their former beauty and richness especially by their interior decoration where wood, mirrors, tiles, stucco and marble prevail. This particular family also had held the *ʿomdiyya*, the office of the village headman. At the time of my research the aged widow of the last ʿOmda from this family was still living with a daughter in this house, supervising the agricultural production of the villagers. She has a son who is an electrical engineer living and working in Cairo. He comes on weekends to visit his mother. He has recently built a large villa just beside this old house, using an entirely modern material – concrete stones. He has also built a well with an electrical pump; he has a unit-furnished kitchen; and a bathroom with bath-tub and 'French' toilet. The premises comprise a very large courtyard dedicated to harvest work. His plans are to lay out a garden surrounding his house which will give shade and fresh, cool air. The estate, too, is walled and it is therefore not easy to see the interior from outside. The son wants his mother and sister to leave the old and uncomfortable house and move in here, while he himself sees the building as a kind of summer resort he can take refuge in from polluted, noisy and overcrowded Cairo. This stressed Cairene businessman is engaged in creating a kind of rural idyll and materializing imaginatory country life by investing revenues from the family land in the village.

Poor people, needless to say, do not have the rich families' power and influence. Their access to space and their ability to create and occupy space consequently are much more restricted. The most obvious instance is of course their lack of agricultural soil. They do not have land, or if so only tiny plots. Moreover, access often is restricted as they only have some kind of usufruct rights to it. Their houses are also small, one storeyed, cramped together and connected by narrow alleys. Lacking the means for internal functional separation they temporarily occupy public space for such chores as cooking, washing up, drying cleaned cereals or the like. They do not have a kitchen garden. There is no separate stable for livestock – humans and animals all live under the same roof. It is for this reason that poverty is often talked about in terms of dirt. This is obvious from the following opinion expressed by a villager from an adjacent community: 'everything is better here and cleaner than in Kafr al-ʿIshra. Nobody here has his *gamusa* (the water buffalo) in his house'.

While one can distinguish space occupied predominantly by the landless and the small peasants from that where richer families live, there are nonetheless no distinct quarters for the poor and the rich. One reason is the steady growing of the residential areas of the village. There are, however, the mental maps. Villagers have a clear concept of such rich and poor quarters: here a dichotomy exists between the territory of poverty, landlessness, illiteracy, dirt, and, on the other hand, the landscape of wealth, education, power. This dualism finds its clearest expression in religious symbolism: power and wealth are connected to the Christians of the village while poverty and illiteracy is reserved for the Muslims. This has a historical origin, as it happened to be a Christian family which decided, as many others did at that time, to establish an estate on the family's fields. And, as Muslims made up the majority in the region and as they had lost their land, like many peasants at that time they had no chance but to secure survival by settling on the landlord's land and by working for him. The only peculiarity was that the landlord happened to be Christian, whereas in many other villages in the region with a similar history he was Muslim, foreign, a state official, a banker. These Christian families have, by and large, been able to retain significant influence until today to the effect that political and economic power is still in their hands. The territory of the poor is often called clandestinely *harat al-ʿabid*, the 'quarter of the slaves' or 'black'. The most blunt definition of this sort once came from a sister of the ʿOmda when she accidentally heard that I was about to visit some Muslim families: 'Oh, don't you go to visit them, they are so hostile towards us, don't you know that they are only Muslim. They won't treat you in a gentle way and you will certainly catch lice there!' A similar incident took place on the occasion of the 'night of henna' celebrations the evening before marriage. When mentioning to some Christian girls that I wanted to join the festivities they said that it would be better if I didn't; in any case they wouldn't because the Muslim *shabab*, 'young men' were not good. They would not behave in a correct way with young girls. They even mentioned some names. The (rich) Christian families, in turn, are sometimes referred to as the *gamaʾat al-maʿallamin*, the 'community of the patrons' or the 'landowners'. These 'mental maps' are also detectable in the built-up environment. Muslim houses are often decorated by a circle made of straw, symbolizing the Muslim crescent, and Koranic verses are also painted on the walls; Christian houses are marked by a cross.

Then there are the most powerful symbols: the church and the mosque.

The symbolic representation of this ongoing struggle over the occupation of space has also a temporal dimension which again finds its clearest expression in religious terms. In the following example I want to show how in a very literal sense *Zeiträume* (time-spaces) are created and occupied. My first night's sleep in the village was rather brutally brought to an end some time before dawn when the mosque's loudspeaker started to transmit the religious morning programme of Radio Cairo. A male voice gave a lecture on religious instructions for about half an hour and it did so at such a high volume that there was no thought of sleep anymore. This was eventually followed by the habitual call for prayer. I felt terrorized. I soon had to learn that this was no exceptional event and it took me some weeks to get used to these nightly religious lectures. The Muslims, who, during daytime definitely have to play the subordinate role in the village community, at nighttime, with the help of electricity and loudspeakers, manage (in the form of its religious programme) to get the state on their side. As this concerns the religious duties of the Muslims, who are dominant in Egypt, nobody dares to protest[7] – a show of power by the Muslims in a village where they are normally the underdogs. A similar occasion could be observed during the night of henna of a Muslim girl, at a date which happened to coincide with the fortieth-day memorials of the death of a very influential Christian landowner. Both houses are situated only 100 metres apart. While in one house the black-dressed women were gathering, in front of the second house neon bulbs were placed in position and loudspeakers were playing Egyptian pop songs very loudly. The scene was made even more clamorous with the villagers coming to congratulate the Muslim girl.

I want to cite another example showing the Christians' cyclical occupation of 'temporal space'. During the yearly *usbuʿ an-nahda*,[8] the commemoration of Mari Girgis (see Chapter 3) the church's patron, Christians decorate their streets with coloured pictures of the saint. This is the most important annual event in the Christian community's religious calendar, celebrated during a whole week at the village church with prayers and ceremonies. Many Christians from the surrounding villages and towns, and even from as far as the governorate's capital, come to participate in the divine services. It is at this time that the offspring of the old absentee aristocracy can be found in the village. Their women and children arrive in fashionable dresses, their hair carefully curled, and wearing heavy make-up. During the celebrations of the *usbuʿ an-nahda*, the old landowning aristocracy attends in such great numbers that Christian villagers themselves can hardly find a place in the church and are therefore often forced to follow sermons from the churchyard, sitting on the ground and holding their children in their arms. However, this 'internal' Christian occupation of space remains enclosed within the church's walls. While in mundane everyday life Christian peasant families are conscious about the fact that they are to a large extent toiling to make the landowner's profits, in the context of these religious ceremonies, and before the 'excluded' Muslim population, the aristocracy are a symbol of their own superiority: Christian peasants, no matter how poor they may be, enter the church together with the absentees and jointly participate in the festivities, whereas the Muslims are excluded and have to stay outside.

On the preceding pages I have tried to delineate Kafr al-ʿIshra's social geography, starting with the observation that village stratification finds its expression not only in sign systems which villagers can easily read from the built environment but also from action. Hence space and time appear as important categories intimately connected to power and wealth. This is so especially in the agricultural village where an household's wealth, power and prestige is to a very large extent derived from the amount of land possession. Large landholding families are also socially and politically dominant in the community. Thus the degree to which a family is able to control and create space is a function of its status in village society. Yet such sign systems do not only consist of elements attached to the built environment. *Zeiträume*, periods of time, are equally contested over and give us an idea that such systems are by no means static but, rather, are essentially dynamic. The regular compulsory early morning religious instruction has given us an idea of this periodicity. Here, another 'modern' device comes in. Electricity is made use of in the same way, which in turn has made possible the use of loudspeakers and the transmissions of religious programmes of the state. Thus, vociferousness has also become a device to create and defend territory. All these are elements in a kind of 'everyday life civil war' (Stauth 1990d: 356) going on in the village. Religious affiliation and religious symbolism clearly are most important as identity markers in this respect. However, bringing in the Muslim/

Christian dichotomy one has to be careful about what one is describing when referring to such an everyday-life civil war. And villagers would certainly reject this picture as, at best, too blurred to describe their religious reality. They are certainly right to do so when one takes into account militant religious fundamentalism in Egypt. The sort of atrocious confrontations characterizing the scene, especially in the region of Asyut, does not take place here. Muslims and Christians in the village get along very well with each other: a Christian woman with a lack in milk had her baby nursed by a Muslim woman – this is only one significant expression of peaceful coexistence. And I have also seen an old Muslim widow decorating her home with the pictures of Christian saints. Both parties frequently stress that they live together in harmony. Without any doubt, too much would be at stake if this kind of everyday-life civil war, which is fought with symbolic weapons, emerged into any kind of open confrontation. Villagers, of course, are aware of that. As a matter of fact, there are no religious political undercover activists, no members of Muslim or Christian fundamentalist organizations in the village – which might be the most important reason behind peaceful coexistence. In the analysis we therefore have to be very precise when speaking about religious stigmatization. We have to be explicit about whom we are talking, and also look for the inherent aims followed. What is meant by 'the Christians' stigmatizing 'the Muslims'? Connected to this question, and equally important, is the fact that what I have presented so far is not some kind of play enacted on a village stage disconnected from the world outside. This picture calls for the taking account of historicity and, equally crucial, of the community's being an integral part of the wider world, of national and global networks. If some villagers present this sort of picture about Islam and Christianity we have to try to find out who does so and to understand the underlying motives and aims.

AGRICULTURAL PRODUCTION

So far we have dealt with symbolic aspects of power, wealth, and respectively, poverty and powerlessness, and we have in a way left aside the material basis of this symbolic expression. It is this issue I want to deal with now. From the very beginnings of the village until today this material basis of wealth and poverty has been agricultural land and access to it. In recent years much has been said about the declining importance of agricultural production in the Egyptian countryside to the benefit of off-farm employment. Many educated young people take up government jobs which secure them a minimum monetary income. Often they shift to 'part-time farming', working in the government office in the morning and going to the fields in the afternoon. Another much-debated question is the abandoning of agricultural production by migrants. The decreasing importance of agriculture in the securing of one's living is clearly observable from one of the urbanized villages in Minufiyya we did research in. Here, 81.7 per cent of the household heads who worked indicated a job outside agriculture with only 20.9 per cent still having access to land – figures which prompt us to reconsider concepts of the village. Yet, such tendencies are of minor importance in the village of Kafr al-ʿIshra. Here, agriculture still plays a very central role in the lives of most of its families as 65 per cent of the heads of household work in agriculture and 64.1 per cent have access to land. Hence, human life is still to a large

Plate 4.3 Women planting rice. This is women's work – obviously! However, to do justice to the man in the background he has taken off his *gallabeyya* and is distributing the rice plants on the field so that the women can more easily reach them. This is a man's job.

extent geared to the seasonal agricultural cycle and its needs. Agriculture is the most important factor to secure reproduction and the main source of subsistence products: agriculture provides jobs, cash and products consumed in the household.

Access to land

I have already explained in the introduction to this book that agricultural holdings in Egypt are rather small. In this respect Kafr al-ʿIshra with 82.6 per cent of holdings not exceeding 5 *faddan* is typical. Yet compared to the national average (0.9 *faddan*: 0.37 ha) the average holding in Kafr al-ʿIshra is considerably larger with a mean size of 3.26 *faddan*. We may interpret this relatively large size of holdings as an indicator of the general importance of agricultural production in Kafr al-ʿIshra. As off-farm employment is hard to find, and as making a living exclusively from agricultural (daily) wage labour is hardly possible, access to land is a very essential concern. The majority of holdings in Kafr al-ʿIshra (64.7 per cent) comprise 1 to 5 *faddan* (see Table 4.1). In literature on the subject, holdings of up to 5 *faddan* are usually put together to form one group, then categorized as smallholdings. While this is evidently the case, in comparison to holdings comprising tens of *faddan*, such a procedure obscures too much of the socioeconomic reality in Kafr al-ʿIshra. Here, acreages comprising 3 to 5 *faddan* already place the holder in the group of the affluent who dispose of a relatively secure and sufficient base of income. But even possessing only 1 *faddan* remains a dream and goal of many households. Though such

Table 4.1 Distribution of owned and rented land

Size of plot (faddan)	*Number of plots*	*Percentage**	*Percentage*†
< 1	33	11.5	17.9
1–3	89	31.0	48.4
3–5	30	10.5	16.3
5–10	18	6.3	9.8
> 10	14	4.9	7.6
No land	103	35.9	missing
Total	287	100.0	

Source: Village survey 1987/8.
Notes: * Percentage distribution of access to land with regard to all village households.
† Percentage distribution comprising those village households with access to land.

acreages are not sufficient to cover all the needs of the household, income and subsistence products derived from such plots nevertheless form a vital and secure basis for the household's survival. Only 7.6 per cent are 'large landowners' in Kafr al-ʿIshra with holdings surpassing 10 *faddan*. The largest landownings in the village are one of 32 *faddan* belonging to a descendant of the old absentee class, and one of 22 *faddan*. The latter belongs to the village headman. The ʿOmda, however, owns more land in other villages, and this is not registered at Kafr al-ʿIshra's cooperative.

Rental systems

In my description I have used the term 'access to land' rather than 'landownership' – that is, land both owned and rented. This clarification is of considerable importance as a substantial portion of the village land, 189.4 *faddan* (39 per cent) out of a total of 486 *faddan*, is not cultivated by its owners but by tenants. While this ratio of rented land is already considerably large the actual proportion of the village land cultivated in rental agreements is probably much higher. Villagers I asked estimated that about three-quarters of the village land was not farmed by its owners. The reason for this discrepancy is that the official proportion is taken from official cooperative data where only legal rent arrangements are registered. However, widespread illegal practices have created a situation where land is leased but not registered. While households with smallholdings also occasionally rent out land, the prospective renter must usually turn to the large landowners. The size of these rented plots is relatively small, as 89.9 per cent (n = 89) of them are below 3 *faddan*. Acreages of this size – which are largely due to the predominant rent system described below – barely secure the survival of the tenant family. Of the remainder, according to the village survey of 1987/8, 7.1 per cent (n = 7) rented 3–5 *faddan* and 3 per cent (n = 3) rented over 5 *faddan*. As elsewhere in Egypt fragmentation of rented and owned land into small, separately operated plots is an important reality. Also, the relatively larger property of the big landowners is fragmented into small units of production, operated under rental agreements with different smallholders. Another side-effect of this system is that large landowners thus establish networks of tenant dependency.

In the light of landlessness and difficulties to find a stable off-farm job, access to land is of extremely high importance to the small and landless peasant households. Crucial to this are the three most important forms of rent, which I shall now describe.

There are principally two legal forms of tenancy practised in Egypt today, both introduced and sanctioned by the Nasserist land reforms of the 1950s and 1960s. At that time the smallholder's position was consolidated through legal protection: land is registered at the cooperative under the name of the holder and the landowner cannot easily evict the cultivator from his land. In the case of the holder's death the right to cultivation of the soil is inherited by his family.[9] In Kafr al-ʿIshra the predominant rent is a form of share-cropping called *muzaraʿa bin-nuss*,[10] with inputs and outputs being divided between the partners. While the harvest is shared on an equal basis, it is the landowner's job to provide the soil, whereas the cultivator contributes the necessary labour power. The cultivator is responsible for all the working tasks in the field such as the preparation of the soil, the sowing or planting, irrigation, the harvest, etc. Costs for fertilizers, herbicides, seeds and taxes or for the transportation of the harvest to the market are shared in equal parts between owner and cultivator.

The other form of permanent rent is a cash rent which is simply called *igar* ('rent') in the village. It is, however, only in rare cases to be found there. Cash rents were restricted by the Nasserist stipulations to seven times the basic land tax. As the government did not substantially raise land taxes, rents have also consequently gone up very slowly. The rent of 1 *faddan* was around £E60 according to the quality of its soil, and only in 1989 was it to reach about £E200 a year. Holders of *igar* land are thus definitely privileged and it is no wonder that today no landowner would enter into such a tenancy agreement. The fact that in Kafr al-ʿIshra it is share cropping which is overwhelmingly practised may be another reason for the village's general poverty. Both of these legal forms of rent agreements are contracted on a permanent basis, to the effect that the owner cannot evict the peasant from the soil without the latter's agreement unless he has failed in its cultivation. Stauth (1990d) remarks that such tenancy is usually engaged in with poor members of the big landowning families whose loyalty to the landowners is thus assured.

For landowners it has thus often become more profitable to sell the land and invest in real estate in urban areas, an opinion expressed by a man in the village: 'the *iqtaʿiyyin* [the 'feudalists'] do not have any interest in the land anymore. They sell their soil to the migrants. If, for example, they want to dispose of 2 *faddan*, they sell them to the migrants for £E30,000, and with this money they build a house in Sinbillawein with six flats, each of which they rent out for *igar*.' Worse for the landowner, he or she cannot sell the land unless the cultivator agrees to it and he is then entitled to half the land.

Households without permanent access to land often have to resort to a form of illegal seasonal rent. Such rental agreements are not contracted on a permanent basis, but are usually restricted to one agricultural cropping period: seventy days for maize or six months for clover. The *fallah* has no right to claim a second term of cultivation if the landowner is not willing to comply. This kind of arrangement is also far more expensive: a *faddan* destined for the cultivation of *barsim*, at the time of field research cost about £E400 to 500 depending on the quality of the soil – a sum which often has to be paid in advance before taking over the land. As this form of agreement is so expensive, renting more than 1 or 2 *qirat* is out of the question for most households. These plots are rented with the sole purpose of subsistence production (see also Glavanis and Glavanis 1983: 55). Usually clover is grown to feed the household's animals, or maize which serves as a basic food staple for both cattle and human beings. Such rental agreements are in fact quite common in the village. There are also some rich 'entrepreneurial' households from outside the village renting land, a recent development much more common since the implementation of *infitah*. They rent large areas and plant them with profitable new cash crops, especially vegetables and fruit. These crops are now finding new and rapidly expanding urban markets. Profits here are high, if compared to the traditional crops, as prices are subjected to market forces and not fixed by the government.[11] When I came back to Kafr al-ʿIshra in June 1992, an entrepreneur from outside the village had rented 35 *faddan* to grow tomatoes, and the price per *faddan* was £E1,000. It is here that the discrepancy between legal and illegal *igar* becomes most apparent.

Transferring wealth outside the village

Certainly the positive aspect of the Nasserist system of agricultural production has been that the *muzariʿ*, the cultivator, had a guaranteed access to land. In this way at least part of the reproduction of his household was secured. This definitely has been a positive aspect of land reforms. On the other hand the actual amount of the output at the disposal of the cultivator may be often very meagre if mandatory government regulations, a bad harvest, and share-cropping coincide. The cotton harvest of 1988 has been one example. Growing cotton, the 'government crop', has generally not been very rewarding for peasant households in recent years because of heavy indirect taxation, relatively high labour input and exhaustion of soil. According to Richards (1991: 61) 'yields fell by over 20 per cent and production declined by roughly the same amount from 1984 to 1988' in Egypt. In the year of the field research cotton harvest brought extremely poor results with only 3 to 4 *qantar*[12] instead of an average of about 8 to 10 per *faddan*. Here is the assessment of a woman from Kafr al-ʿIshra who also had planted cotton:

> The *hukuma* ['government'] pays us this year £E100 for each *qantar*. The harvest of this year has been so poor, that we will not get more than about £E300 to 400, and from this the cooperative's advances in seed, fertilizer and pesticides still have to be deducted. This will cost us about £E150. The costs for wage labour and irrigation have still to be deducted from the rest. If you do not own the land, but if it is *muzaraʿa bin-nuss* land, then you will have a net income of about £E75 to 125 – and this is the result of six months of work! And we achieve this result only if we depend on our own labour and if we *mnitzamil* [this is the expression for mutual, unremunerated labour shared between different households] and do not hire *anfar* [daily wage labourers]. You know that there are some people who say they did not earn anything from cotton this year, they have worked entirely for the landowner and the *hukuma*. You now certainly understand why nobody, not the peasants nor the landowners, are interested in growing cotton!

This statement mirrors the fact that beside the large landowners in the village the central government is to a similar degree engaged in the transfer of wealth to exterior markets. In this sense Mitchell has argued that the state, through its mandatory growing system, price policies, subsidies and compulsory delivery, has syphoned off a large proportion of the rural surplus value. Quoting Dethier (1989) he states that 'the net effect of government policies between 1960 and 1985 was to appropriate 35 percent of agricultural GDP' (Mitchell 1991: 26).

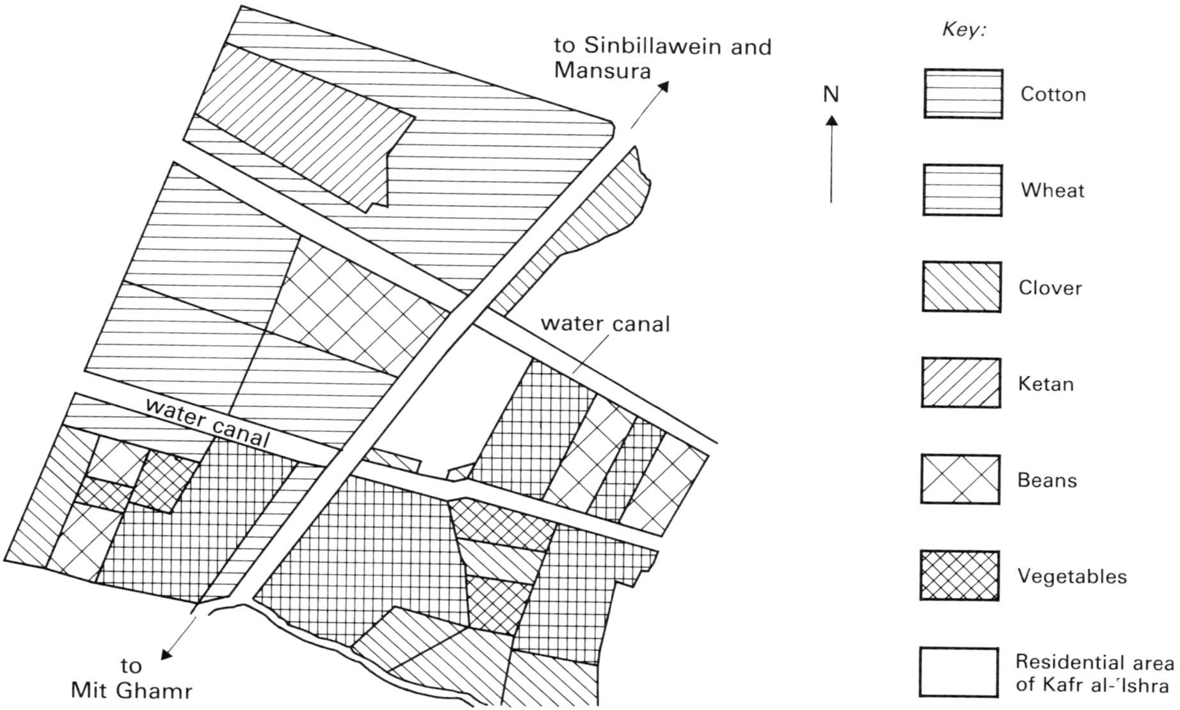

Map 4.2 Mandatory growing system in Kafr al-ʿIshra for the winter period, 1988/9

The cultivator's dependency on exterior interests and regulations, as well as the extraction of surplus, is to a large extent due to the already mentioned mandatory growing sytem which regulates agricultural production according to government directives. On the village level it is the agricultural cooperative, a government agency, which surveys the cultivator's compliance with these regulations. Map 4.2, similar to the one to be found in the cooperative's office, illustrates this mandatory system for the agricultural year 1988/9. It shows the parts of the village land earmarked for special crops. Peasants are not free to cultivate their land according to their own interests and directives, but they are told what they have to grow on their fields. There are, of course, numerous ways to circumvent these government stipulations. Among these are the exchange of plots among the peasants or their ignorance of the regulations. Those who violate the regulations are supposed to be fined. It is obvious that offences hold less fear for the well-off who can afford the money to pay fines. I noticed this with regard to the production of rice, one of the basic food items consumed in the village. Because of water shortages the growing of rice was forbidden in 1988. One medium-large landowner disregarded the regulation. He should have been fined but this never happened. There are thus always methods to circumvent the mandatory growing system, and these are open to the big landowners especially (see Adams 1986: 68–9). But even if they comply with the regulations they have better chances for good results. Here is an example. Asking a cooperative employee for the reasons why one of the largest landowners had been particularly successful the old stereotypes about the peasantry were forwarded:

> He does what we tell him. He is very intelligent. When we tell him 'irrigate' he irrigates. When we tell him to use a certain quantity of fertilizers he does that. But the peasants don't do what we tell them. They think they know everything better than we do. They are too stupid.
>
> (A cooperative employee)

Later a peasant commented on the same issue in a quite different way:

> His cotton harvest was the only one with good results. While everybody had only 1 to 3 *qantar* he had 8 to 10 per *faddan*. You ask me why? Because his relationship with the fellows in

> the cooperative is very good. He invites them for tea and I do not know for what else. Therefore they do everything he wants. They give him more fertilizers and pesticides.

On the map the areas of the village land earmarked for cotton in summer 1989 are specified. Regulations ensure that cotton is not grown on the same plots as in the previous year, which would damage the fertility of the soil. On the other hand it is interesting to note that 40.5 per cent of the village land is reserved for the obligatory cultivation of cotton. The cooperative keeps a register of all holdings, the name of the holder, the size of the holding, as well as the kind of access to the land. On the basis of these records, seeds, fertilizers and pesticides are distributed in advance and have to be paid for after the harvest. The cooperative also receives the output of certain crops. Cotton, for example, is to be obligatorily rendered to the government, as well as half of the rice harvest, while maize may be marketed through the cooperative at a fixed price.

The peasant household's agricultural production is thus to a large extent subject to the government's regulations. This system, by which the cooperatives specify and enforce crop rotation, overlook the distribution of inputs, and secure the delivery of a substantial part of the crop at fixed prices, has the effect that implicit taxes and subsidies transfer resources out of agriculture (Richards and Martin 1983: 5). Though holdings are small and often further fragmented into various plots which are sometimes even situated in different parts of the village, for the government the benefits of a large plantation are secured through the mandatory growing system: under its regulations the land is planted with the same crop, cultivated and harvested at the same time, while only the input of labour power is derived from various households.

NON-AGRICULTURAL INCOME-GENERATING ACTIVITIES

Walking through the narrow alleys of the village an attentive visitor will discover that different forms of petty trade belong to the most important forms of making a living outside agricultural production. Here and there one will find a petty trader's enterprise, established by a woman who has consecrated a room or the veranda of her house to store and sell merchandise. There are several women in the village housewives can count on for always having a stock of the most important vegetables like tomatoes, eggplant or *gargir* (a sort of fresh herb used for salad) for sale. Occasionally other women engage in petty vegetable trade in the village, if on a far more restricted scale. Whether they sell or not often depends on whether they have in their pockets the necessary Egyptian pounds to buy a box of tomatoes or potatoes from some merchant in Sinbillawein. These women also close down their enterprise during agricultural peak seasons because then the wages they can earn are higher than the returns of their trade. Thus enterprises open and close down frequently and irregularly. Besides lack of cash women face logistical problems. They need a daughter, a daughter-in-law, a son or their husband to get them their merchandise from the district town or to look after their children when they are away. Selling vegetables is not as easy a job as one is inclined to suppose. Women get up around 4 a.m to go and fetch the merchandise and only return around 10 a.m. Often they carry over 100 kilos in boxes and sacks, using the public transport pick-up as any other sort of transport would be too expensive. Profit margins are also extremely limited. Another observation is that this kind of business is still subject to the barter system. I watched about one-fifth of the women exchanging rice or wheat for vegetables.

Another important opportunity to get hold of cash is to sell home-produced goods – also a domain of women. Most important in this respect are milk, cheese and eggs. Some women take their produce to the market to sell it on their own, while others hand it over to a trader. Women known to sell are also directly approached by villagers who are in need of something. The selling of home-produced items is, however, an ambivalent issue. On the one hand, women – and perhaps their husbands even more than them – are proud of home products as they are generally regarded as of high value. The amount of subsistence products disposable symbolizes a household's ability to provide healthy and tasty food for its members and this should only be given away if there is more than needed. Selling could therefore be interpreted as a sign of affluence. Yet, on the other hand, I always had the impression that the selling of such products, be it on the market or to a neighbour, was somehow concealed as it is also interpreted as an indicator of poverty: that the most precious goods of the household, essentially those to be given to the children, must be sold in order to get hold of cash. Anyway, selling home-produced goods is a possible money earner in crisis situations though amounts will not be very high.

And petty trade is one of the few possibilities open to the poorest women in the village, such as those widowed or divorced, and who have no land and no other persons or families caring for them. Itinerant women traders from other districts occasionally complete the scene, offering grape or guava from baskets they carry on their heads.

Three shops in the centre of the village cater for such 'luxury' consumption needs as ice-cooled Coca-Cola or Fanta, and ice-cream mainly sold in small cornets to children. Housewives usually patronize these shops in emergency situations, such as if they are running out of the monthly purchased subsidized food items or out of their own stock of subsistence products. Then they come to buy Egyptian-made macaroni and canned tomato paste. The range of commodities mainly comprises canned sardines, steel wool, electric bulbs, soap and oil. One of these shops also functions as a post office, selling envelopes and stamps while at the same time collecting and distributing letters. There are some other, very tiny '*butiks*' (wooden sheds) where, besides the inevitable soft drinks, Marlboro and Aspirin are sold, as well as hubble-bubble tobacco, yeast and sweets.

There are also a number of manufacturing enterprises, *warsha-s*, nearly all of them situated in the basement of houses along the asphalt road, turning it into a kind of 'small industry' zone. There is a small workshop repairing the tyres of cars or those of agricultural tractors, and a welding workshop. One of the larger enterprises is a carpenter who produces modern, urban-style furniture. His services are predominantly requested by young couples about to marry. A *mabyada* where rice is husked can also be found here. There is yet another successful village entrepreneur who repairs the rubber tubes of diesel-operated water pumps which have recently made their appearance in agricultural production. All of these enterprises are so successful that they even manage to employ one or two workers who, though paid on a daily basis, are quite regularly to be found working here. Situated along the main road these enterprises are also patronized by strangers from other villages or districts who have no difficulty finding them because of their prominent location. These workshops respond to the newly arising needs of the countryside: the villagers are becoming more and more motorized, and machinery is increasingly introduced into agricultural production. It is no surprise that those who get a job here consider themselves to be lucky as they can count on a relatively regular daily cash income of about £E3–5.

Off-farm employment within the village remains restricted, by and large, to these small enterprises. There are other jobs occasionally to be found in connection with the gradually emerging mechanization of agriculture. Here labour power is needed to feed the machinery, cutting rice and wheat straw or bundling it. To date, five of the larger landowners have invested money into such machinery and these kinds of jobs consequently remain rare. With a daily wage of about £E3, income from working these machines is slightly higher than from agricultural labour, which remains the most important opportunity of cash income for many families. According to the season and the kind of job, wages are as low as £E1.5 to 2.5. Here are a couple of examples for wages paid in 1988 in Kafr al-ʿIshra: £E1.5 a day for picking tomatoes, a job usually performed by the *ayal*, unmarried young women and children; £E1.5 a day for picking cotton, a decline from the 1987 rate of £E2 a day, although payment went up to £E2 later in the season when the cotton harvest coincided with that of maize thus leading to a shortage of manower; the harvesting of maize was paid at £E2.5.

Due to the ongoing building boom the construction sector has also become a source of jobs and there is a market for daily wage labour in the district town. Men who do not find a job inside the village often turn up here in the early morning hours waiting for employers to pass by in their cars and to pick them up. As a matter of fact, only the young and strong have a chance here. The more fortunate have established personal relationships with employers who call them or to whom they can address themselves directly without going to the 'slave-market'. Usually these persons are privileged because they have certain skills such as plastering or concrete mixing. Those who have profited from the Nasserist system of education, who hold degrees and have found jobs in the government's bureaucracy, are among the ones who daily leave the village to attend their working place in a neighbouring village or the district town: these are teachers, the personnel of cooperatives, social units and public administration. Though a regular and stable income is considered to be a source of security, the average government employee's monthly salary of £E40–45 is too low to guarantee a sufficient base of income for the household.

CONCLUSIONS

Many facts contribute to the assessment that agricultural production in Kafr al-ʿIshra at the end of the 1980s is still organized in the *ʿizba* form, where cultivators have access to a plot of land on which subsistence production is relatively autonomously organized while at the same time cash crops and labour power are exported to exterior markets. In this sense I have dealt with such features as access to land, legal and illegal rental systems, the mandatory growing system, the dependency on the large landowners and the lack of employment opportunities outside agriculture. All these facts contribute to the 'stability of instability' whereby an existence very near the subsistence line is perpetuated. The labour force of these households is not fully proletarianized, nor is cash-crop production organized predominantly on large plantations. Within this framework these households remain free to allot labour power and to dispose of those parts of the harvest which are not obligatorily to be rendered to the state cooperative or to the landowners. The big landowners in the village profit[13] from the peasant households, as cheap labour force remains one of the major economic assets of the village community exploitable to their interests. As the small peasant household's subsistence basis is so scarce and as they are invariably bound in crop-sharing arrangements, as off-farm employment is rare, they consequently function as a pool of cheap labour power to be flexibly employed in cash-crop production on the big landowner's fields. The fact that in the village of Kafr al-ʿIshra mechanization of agriculture has not made great strides may well be an expression of this. Today, besides the descendants of the old absentee landlords, there is a new indigenous, village-based stratum of medium and big landowners who are able to exploit the cultivators's labour power to their own profit. And here, the ʿOmda is perhaps the most prominent example. But there is also an exterior force, the state, which through cropping regulations and low-pricing policies transfers values outside the village. In fact, in the wake of Nasserist reforms the Egyptian countryside had become a 'state domain' – one large *ʿizba* operated primarily according to the state's interests (Richards 1982: 181, Stauth 1981: 72). In this sense Stauth argued, that

> the *'izba*-system remained the form of keeping the small peasantry alive by inverting their labor force to a system of internationally articulated commodity production and by

'externalizing' the networks for securing and distributing the necessary means of susbistence.

(Stauth 1990d: 123–4)

While from a structural perspective the small peasant household's function for external forces is thus secured, life is a permanent struggle to prevent a deterioration in its economic situation with the risk to fall below the 'subsistence' or 'poverty line' ever present. It is a struggle which includes a permanent search for additional sources of income: wage labour in agriculture inside the village or further afield, and also offering on the market what has been produced in the household. In a situation of powerlessness, of lacking alternative structurally transmitted opportunities for income generation or access to land, it is the household's struggle for survival that proves vital for the perpetuation of this system. Now, as in the middle of the last century, the disciplining of the work-force and its linking to the system is comparatively easy to organize: on the *ʿizba* the *taʿmaliyya* were granted a subsistence plot by the landowner and they thus enjoyed a status preferable to that of the *tarahil* who were totally dependent on wage labour – the threat to lose their position by not fulfilling the landowner's expectations could always serve as a disciplinary force. Nasserist reforms have enhanced the cultivator's position by small land grants and by legally securing tenure systems in his favour – at the same time the reforms thus had the effect of securing the cultivator's attachment to the land. Today it is the share-cropping system, as well as the obligatory delivery of (part of) the crop and the agricultural pricing policy, which serves the same aim: as the peasant household must be interested in the maximization of its own crop returns it is at the same time working to augment the returns of the landowner and the government.[14]

Small peasant producers, the village-based landowners and the state have, if for different reasons, an interest in maintaining the *ʿizba* system of production, and I have pointed to the economic and regulatory mechanisms securing its perpetuation. But the system is also maintained by other means, and thus we return to the symbolic struggle over the occupation of space and time and to the construction of the self and the other. Stauth, with regard to the symbolic means of the moral defence of life-styles, maintains that 'the basic mechanisms with which the big landholding trusts of the *ʿizba* system interfere in the moral milieux of use-value production by the small peasantry and their life-styles are the traditions of the community (*taqlid al-qarya*)' (Stauth 1989: 136). In this respect I pointed to the large landowner who had founded the village and who also became its spiritual head, a position which until today is occupied by his family. I have also shown in this chapter that before active interference comes self-styling: the construction of the self and the other. Though today the ʿOmda ultimately derives his authority from the government, an outside institution, his self-styling predominantly presents another image. This is the ideal type of the benevolent big landowner, the patriarch who protects the needy of his community. He has made himself the very symbol of the safeguarder of the *taqlid al-qarya* and it is his actions, to no minor degree, which define the obligations an individual has to perform within this moral universe: by sheltering the destitute couple on his premises, by distributing crop rations after harvest, and by sending tissue material on festivities to the needy the ʿOmda signals that he knows his obligations as the landowning patriarch and thereby confirms his legitimacy as the dominant political force in the village.

This importance of a 'traditional' legitimation over a 'modern' one is underlined by the following example. Talking about conflict-resolving institutions in the village, more than one woman conveyed to me that villagers usually address themselves to George, one of the large, village-based landowners (introduced previously in this chapter) who in fact quite often was described to me as a very benevolent, just, and good-hearted elderly man. Such descriptions were sometimes concluded by the observation that people would hesitate to address the ʿOmda with their disputes and conflicts 'because he immediately tells the police in the district town'. The explicit reference to the community's moral code is therefore not restricted to the village-based large landowners who are dependent on the cultivators' work-force and hence their loyalty. And symbols of legitimacy must be understood to be so by those addressed. We can in fact observe that the 'clients' definitely have the power – and an interest– to call in this kind of charity and protection and thereby make sure that the styling of the self also has practical and factual aspects concerning their own profit. It seems to me that both economic forces – the large landowners and the dependent poor – join forces in the construction and maintenance of the *taqlid al-qarya* as an asset which they use for different purposes. Both are reconstructing, reinventing the *taqlid al-qarya* in a permanent process. The village-based landowners' interests in the perpetuation of the system are evident as they are dependent on the village's resources – that is, on agricultural produce, labour power and the loyalty of the cultivators. But this is equally the case for large sections of the community's poor stratum: given the scarcity of off-farm employment opportunities and its inadequacy to form a sufficient source of income, given also the almost non-existent social security, they are thrown back on the community in their struggle to survive. They remain dependent on the land, the jobs, the cereals, and the mediation the large landowner can provide them with.

It is at this point that religion comes in again. Claiming to be the patron of the whole community, the ʿOmda himself cannot allow himself to treat Christians and Muslims differently. It is therefore not surprising that he proposed both a Christian and a Muslim as my research assistants – Samia and the *mutadayyina* (see Chapter 3). I never observed him (or his wife) join in other Christians' stigmatization of the Muslims. For a while I asked myself why it was especially the small, landless, or near landless peasants to whom this religious stigmatization seemed pivotal – and not the large landowners of the village. Given the unequal distribution of power and wealth in the village why did these poor villagers not choose their enemies in class terms? Why was it those different only in their religious beliefs, but suffering from the very same situation of deficiency, who were 'hated' and constructed as the other? Rather than seeing the enemy in the big landowners or the state, those belonging to the same social stratum, although of different religions, are represented as what they actually are – contenders over scarce resources. The focus then turns to factors which could legitimize one's claims as more justified than that of the rivals in the competition over access to those who possess land, power, influence and connections. Here, being a poor member of the landowner's family serves as such a vehicle, as does belonging to the wider 'family' of the same religious community: this latter may very well be brought to the fore to justify priority rights in the allocation of the patriarch's benevolence and obligations. But this is also relevant with regard to outside job opportunities: I have come across several young men who found a job through Christian mediation, mothers addressing themselves to the priest who found them employment in Christian-owned factories in Cairo.

In its struggle to secure reproduction the poor peasant household has to fend for itself to get hold of resources. Naturally, this plight is not restricted to the normative idealization of the *taqlid al-qarya* and its symbolic defence. Looking for other strategies immediately leads us back to the realm of economy. It is here that migration as a structurally given opportunity has become of major importance as a possible tactic in the household's autonomous allocation of its labour force. It is not difficult to imagine that migration, both as a way to break through the village boundaries and as a way to search for external sources, may be looked upon as a potential solution to the household's economic crisis. It is a possible way out of the dilemma of the limited base of subsistence production, of the scarcity of wage labour, and can be seen as a means of escape from dependence on the landowner. While there are many motivations to migrate, it is in these poorer quarters of the village that migration gets a special significance within the households' strategies to secure reproduction. Can this assumption of the functionality of migration for the household be corroborated by empirical data? Before approaching this question, we will next try to assess the dimensions of out migration on the village level.

5

KAFR AL-ʿISHRA'S PEASANT HOUSEHOLDS

Production sites of labour power for international markets

I have been arguing that agrarian production in contemporary Egypt displays basic characteristics of the *ʿizba* system of production: the peasant household still organizes reproduction autonomously to a considerable extent while at the same time a large part of its agricultural produce is diverted to exterior markets. The explosion of oil prices as an event on the macro-level has contributed a new element to the *ʿizba* system of production, as today, besides cash-crop, another commodity of major importance is thrown onto international markets: migratory labour power. This has principally been my line of argument in Chapter 2, where I have treated this point from a macro perspective.

In this chapter I will turn to the place of production of this labour force: the agrarian village of the Third World and its households. Returning to Kafr al-ʿIshra, I shall describe the main features of this migratory labour power by consulting statistical data collected during field research in 1988 (see Chapter 1, note 7). This will give us an understanding of the importance of out-migration from the village and its basic characteristics, which in turn will permit us to realize the ways in which this specific labour power responds to world market interests. A second perspective pursued in the latter part of this chapter is the relevance of migration for the subsistence-producing small peasant household. I have already argued that these households are not able to secure the survival of their members by engaging in only one economic activity, but that they have to be very resilient in combining different and ever-changing income-generating activities. I shall therefore investigate, in the second part of this chapter, the ways labour migration articulates with other economic practices by introducing the families of two lanes in one of the poorer Muslim neighbourhoods of Kafr al-ʿIshra.

STATISTICAL FEATURES OF KAFR AL-ʿISHRA'S MIGRANTS

The sharply rising oil prices of the mid-1970s mark a turning point in Egypt's history of migration. From that date onwards out-migration from Egypt as from many other Arab and Asian countries towards the Gulf economies rapidly increased. This trend is also easily recognizable in Kafr al-ʿIshra. However, its migrants were to join in relatively late as until 1978 there had only been one migrant, and in 1979 only two persons left (see Table 5.1). The turning point for Kafr al-ʿIshra was to come as late as 1980, at a time when migration in other villages had already become firmly established. In 1980 sixteen migrants departed and from that time onwards an increasing number of new migrants is reported. Finally, in the summer of 1988 when the survey was carried out, almost every second household (43.2 per cent; n = 124) reported at least one migrant.[1] At the time of the field research thirty-three households in Kafr al-ʿIshra had a migrant currently abroad.

Table 5.1 Number of new migrants leaving each year

Year	*Number*	*Percentage*
1975	1	0.7
1979	2	1.5
1980	16	11.9
1981	12	8.9
1982	17	12.6
1983	20	14.8
1984	20	14.8
1985	23	17.0
1986	11	8.1
1987	12	8.9
1988	1	0.7
Total	135	100.0

Source: Village survey 1987/88.

According to our census material, the most important host countries of Kafr al-ʿIshra's migrants are Iraq (79.9 per cent) and Jordan (11.2 per cent), which together account for over 90 per cent of all migrant destinations (see Table 5.2). Saudi Arabia ranks third with 6 per cent. This distribution is not peculiar to Kafr al-ʿIshra as these three countries have generally played a major role as destinations of migrants from Egypt in the second half of the 1980s.

Table 5.2 Host countries of migrants

Country	*Number of migrants*	*Percentage*
Iraq	107	79.9
Jordan	15	11.2
Saudi Arabia	8	6.0
Libya	2	1.5
Other Arab countries	2	1.5
Total	134	100.0

Source: Village survey 1987/1988
Note: N = 135 migrants missing values: 1.

One reason for the high ranking of Iraq and Jordan is that the rich Gulf countries, among them Saudi Arabia, have, over the past years, increasingly restricted entrance regulations. Villagers looking for a job in these rich economies thus became confronted with what often turned out to be unsurmountable obstacles. If we remember that migration from Kafr al-ʿIshra started relatively late, at a time when these countries were already beginning to introduce restrictions, this may explain a lot. Today getting hold of the necessary papers often proves too difficult, especially as buying a contract is often beyond the financial limits of migrant households in Kafr al-ʿIshra.

Although this describes the situation generally with regard to the more affluent Gulf economies, Saudi Arabia is an exception as illegal work is easier to find there. This is due to the fact that migrants try to cross the border with the alleged aim of performing the pilgrimage and then manage to overstay the permitted periods. Often they work clandestinely before eventually coming to the notice of the Saudi authorities. On the other hand, with often several thousand Egyptian pounds being needed for a contract, access to these countries goes beyond the financial scope of Kafr al-ʿIshra's households.

Given these impediments Iraq and Jordan clearly are an alternative. Both countries were easily accessible initially as no work contracts and entry visas were requested. The reasons may be seen in Iraq's deficit of labour power during and after the war with Iran, as well as the fact that Jordan itself is a labour exporter where the native or Palestinian migration labour force is substituted by Egyptian peasant migrants (see, in this respect, Ibrahim 1982b).[2] Transport costs to Iraq and Jordan are also considerably lower.

Such advantages, then, lead to an obvious migration decision as one return migrant explained: 'I went to Iraq because travel expenses are very low. I also did not need a *kafil* ('sponsor') or a contract. I also went to Iraq because there are many villagers and friends of mine there'.

The last reason mentioned by this return migrant is especially interesting as here the migrant networks are alluded to which have been established over the years binding together persons abroad with their home communities. These dense networks are extremely important for new migrants as they can count upon assistance to solve their initial problems in a foreign country. Experienced migrants instruct the newcomer how to overcome bureaucratic obstacles most effectively, and provide information about transportation and what life is like in the host society. Persons from the same family, the neighbourhood, or at least the village, provide overnight accommodation and present the newcomer to their own employer or facilitate access to the labour market by communicating to them their personal information and experiences. They may even lend money to new migrants to help them through the first weeks. All this provides security in a foreign environment whose rules and structures have still to be understood. Another effect is that these networks obviously lead to a kind of self-perpetuating mechanism. As most of Kafr al-ʿIshra's migrants work in Iraq, it was the networks here which were most efficient, thus prompting prospective migrants lacking alternative options to leave for this country, too. Therefore, well-established networks in Iraq also provide an explanation as to why that country continues to be a migrant destination.

Migration histories and patterns of different villages vary a lot, a fact which became apparent in our research in six villages in different parts of Egypt. One of the causes for this diversity may well be the particular historical coming into being and development of such migrant networks. Often, mere coincidence may also play a role and explain a certain community's 'luck' in being able to send its migrants to one of the prosperous Gulf states. In our field research one of the Minya villages was just such a case. Here, an astonishingly high 62.7 per cent of the migrants worked in Saudi Arabia. The reason behind this was a travel agency in the village recruiting young men for labour migration. In one of the villages in Minufiyya a similar story was reported. Here, a village woman had married a Saudi and subsequently gone to live with him in Saudi Arabia. Her sister, staying behind in the village, organized an informal recruiting office. Her Saudi brother-in-law came to the village once a year and recruited workers. Villagers said that she first took £E100 from each migrant, but later raised the amount to £E300. Still another example is the ʿOmda of the second village in Minufiyya who himself migrated to Saudi Arabia. He took with him thirty-eight other villagers and acted as a broker for a Saudi *kafil.*

Kafr al-ʿIshra had not been as lucky, as no such mediating service exists in the village, though villagers knew that such travelling agencies were to be found in Sinbillawein and in Cairo. Here, perhaps, the lack of personal ties and affiliations comes into play. After all, the aforementioned £E100 or £E300 sound quite moderate if compared to the several thousand Egyptian pounds people have to pay to such travel agencies.

Migration from Kafr al-ʿIshra is informal (Amin and Awny 1985: 241) as peasants generally do not leave through the mediating services of an official or unofficial institution. Here the preparation and realization of travelling is undertaken individually and with the help of the networks. This fact mirrors a general aspect of out-migration from Egypt, often criticized and deplored by those concerned in national development. They argue that under the given circumstances resources which could be tapped upon for the country's development remain unexploited and are wasted. All in all, international migration for employment from Kafr al-ʿIshra is relatively short term, a characteristic which researchers have also reported from other rural areas in Egypt.[3]

Table 5.3 Number of years migrants spent abroad

Years	*Number of migrants*	*Percentage*
< 1	24	23.5
1–2	32	31.4
2–3	26	25.5
3–4	14	13.7
4–5	4	3.9
5–10	2	2.0
>10	–	–
Total	102	100

Source: Village survey 1987/88.

Roughly a quarter (23.5 per cent) of the remigrants from Kafr al-ʿIshra did not even stay abroad for one year (see Table 5.3).

Another 31.4 per cent remained one to two years. Only 5.9 per cent worked abroad for over four years. The mean time returnees spent in a foreign country for work was 1.7 years, counting all travels.[4] Very obviously, the data does not include prolonged sojourns in the host countries, and definitely gives no indication that migrant community formation is imminent. It is no wonder, then, that visits of family members are a very rare exception: only 4 per cent of the migrants from Kafr al-ʿIshra stated that they had received visits from the village, usually from other family members. Visiting family members are invariably young men looking for a job, visits from migrants' wives and children remaining totally out of the question. This latter would be a far too expensive enterprise, eating up a substantial part of the savings which are meant to be spent in the home village. Also, accommodation abroad does not often allow for receiving a woman. And, after all, who would do all the work in the fields, who would look after the cattle and the children if wives travelled abroad to see their husbands? Furthermore, as the men usually do not stay very long, they will soon meet again anyway.

Table 5.4 Occupation of migrants abroad

Occupation	*Number of migrants*	*Percentage*
Unskilled labour	108	84.4
Skilled labour	14	10.9
Peasant	2	1.6
Farmworker	2	1.6
Semi-professional	1	0.8
Prof/techn/manag.	1	0.8
Total	128	100.0

Source: Village survey 1987/88.
Note: Number of missing values: 7.

Short periods of stay lead to a rapid turnover of this migratory labour force. This quick rotation is also caused by the occupational structure of migrants abroad. Nearly all of them (95.3 per cent) do unskilled or skilled labour (see Table 5.4).[5] 'Skilled' labour often turns out to be only modestly so as anybody who has any kind of training will most probably consider himself as skilled (for example, a person possessing a driving licence). A somewhat disillusioning fact is that even though 23.1 per cent of Kafr al-ʿIshra's migrants are graduates from institutions of higher learning they do not find an appropriate job. Only 1.6 per cent worked as semi-professionals or professionals, technicians or managers;[6] it would surely be interesting to investigate into the reasons. My hypothesis would be that impediments are only partly due to the structure of the foreign labour market, but also result from a situation where graduates hold an official diploma while in reality their knowledge and proficiencies are far behind what one might expect from the certificate.

Unskilled and skilled workers are usually prone to insecure and unstable working conditions. Many of them are daily wage labourers who are not guaranteed work for the next day or week. These workers are very much subjected to the employers' rationales, which enable them to 'hire and fire' according to their daily needs. On the other hand, migrants themselves are mobile as they are always looking for a more lucrative job (see Amin and Awny 1985: 242). Migrants without a contract (i.e., many of those working in Iraq and Jordan) are also 'free' to return home whenever they want – which is more difficult in other countries where migrants are bound by a contract and their passport is in the custody of the employer who often refuses to hand out their salaries and identity papers before the end of the contract. As the majority of the migrants from Kafr al-ʿIshra perform a job requiring only minimal skills it is not surprising to find a high percentage of illiterates (68.7 per cent) among them (see Table 5.5).

Table 5.5. Educational level of migrants

Educational Level	*Number of migrants*	*Percentage*
Illiterate (reading and writing)	92	68.7
Completed primary	6	4.5
Completed preparatory	5	3.7
Completed secondary	29	21.6
Completed university	2	1.5
Total	134	100.0

Source: Village survey 1987/88.
Note: Number of missing values: 1.

Many migrants from Kafr al-ʿIshra come back to the village without a definite plan for a future period of migration. They reintegrate into the community and their household. Then, eventually, the idea of a further period of work abroad takes shape and is realized. Thus a woman talks about her husband's migration:

> He left for Iraq for the first time in 1982 together with a group of villagers, but he only stayed about 40 days because he did not find work. Later he again went to Iraq. And since that time he goes to work abroad and only comes back for holidays. The migrants in Iraq usually do not have a fixed contract, but they work as daily labourers. They work for about a year in Iraq and then they return to the village and they stay here as long a they have a mind to stay. Then they leave again. The longest period my husband was in Iraq was for one year, and that was when we built our house.
>
> (Female, 27, completed primary)

A majority (64 per cent) of the migrants from Kafr al-ʿIshra leave only once (see Table 5.6). But there is a considerable number of persons who embark on two, three or even four travels, leading to an average of 1.47 working periods abroad.

Table 5.6 Number of travels abroad

Number of travels	*Number of migrants*
1	64
2	26
3	9
4	1
Total	100

Source: Remigrant sample 1987/88.

Often migrants do not go back to the same working place or even the same country. Many who had initially been to one of the prosperous Gulf states later went to Iraq and eventually came to Jordan, thus mirroring changing structural conditions. It is noticeable that short-term stays are paired with a high probability of a future migration. This also finds expression in a remigrant's wishes for a further migration. Many remigrants, asked about their future plans, expressed the hope to get a chance to go abroad again.

The high probability of a future period of work abroad can also be correlated to the age of the migrants. Many of them are still young and feel that the best of life is still to come and that there is still much to be achieved and realized which needs financing. At the time of their first migration, 84.6 per cent were under 35, and 93.1 per cent were under the age of 40 (see Table 5.7). The largest group (31.5 per cent) were in the 20–24 age bracket. The mean age of migrants from Kafr al-ʿIshra at the time of their first migration is 27.9. From the perspective of the migrants, their households, and the village community, one of the major reasons of migration for work in the Gulf states is that young men thereby earn the money necessary to live up to their role as household heads. However, viewed from the macro-structural perspective, one of the most important features of outmigration from the village is that the segment of Kafr al-ʿIshra's labour force which is most valuable for the global labour market (i.e. young and healthy men) is at the disposal of the world market.

Table 5.7 Age of migrants at time of first migration

Age	*Number of migrants*	*Percentage*
15–19	10	7.7
20–24	41	31.5
25–29	35	26.9
30–34	24	18.5
35–39	11	8.5
40–44	3	2.3
45–49	4	3.1
50–53	2	1.5
Total	130	100

Source: Village survey 1987/88.
Note: Number of missing observations: 5.

One more migrant characteristic is important in the framework of our discussion. This is the migrants' and their families' integration into agriculure. Not surprisingly perhaps, as Kafr al-ʿIshra is still integrated to a large extent into agricultural production, 83 per cent of the migrants live in households with access to agricultural land and/or work themselves in the fields, be it on their own land and/or as day labourers. Fifty-five per cent of the migrant households farm land, 38 per cent of the migrants had been farmers prior to their departure and another 29 per cent had been agricultural wage labourers. From this it can be seen that most migrants come from places where subsistence production is still of vital importance.

As an illustration of what I have presented on the preceding pages in numerical terms I now want to introduce two accounts by migrants about their sojourns abroad.

My first example is Shahhat, a 40-year-old peasant. He lives in an extended household with his wife, one married and two teenage sons, a daughter-in-law, and a 9-year-old daughter. He farms 2 *faddan in the muzara'a bin-nuss* system. He frequently describes his family as belonging to the *ghallaba*, the destitute. Here is the report about his migration to Iraq:

> When I first arrived in Baghdad I went to some people from the village. We were living there about ten of us in one room. The room's rent was ID150 which nobody could afford. Only those who earn a lot stay in a hotel where everything is done for them: the room cleaned and the bed made. If we had done that we would have spent all our money on the hotel room. That is why we all lived together in one room and shared the rent. There are many Egyptians in Baghdad, many of them live in a neighbourhood called *sahat ash-shuhada*, 'the place of the martyrs', and they only leave if they find a job in another place which would make commuting too difficult. I have been working as daily labourer and usually got ID5 per day. There is a certain place called *suq* ('market') in Baghdad where workers and employers meet. Those who did not have a permanent job or any employment went there in the morning to wait for a job. I have also been working for a contractor which meant that I went for a fortnight or a month to other regions of Iraq. I sometimes worked with an Iraqi who removed old door- and window-frames from houses and sold them at a market. Once I was working for another broker. He cheated me and did not give me the salary we agreed upon. He still owes me ID350 – there is no chance to get the money. We had our friends and spent our leisure time together in a café. Among them were Tunisians, Jordanians, Syrians, Palestinians, Iraqis; we did not only sit together with the Egyptians. I like the Palestinians least because they cheat. [He gives an example.] I like the Tunisians most. They are honest people. I very seldom left the quarter I was living in except for work.
>
> (Shahhat, remigrant)

My second example is the experience of a remigrant who had also been to Iraq. While sitting together with me, and his wife and another female neighbour whose husband had also been a migrant, he entertains us with stories about his experiences in Iraq where he had worked several times:

If I got a new chance I would go to work abroad again tomorrow morning. Working abroad is good and fun. [The two women burst into protest and add that both of them had a very hard time alone.] You learn a lot about foreign countries and other people. I worked in Baghdad. In the beginning I worked in a museum belonging to the state, office hours were until the early afternoon. Therefore I took up a second job with a carpenter until sunset. Later I only worked for this carpenter. Then, I worked in a factory producing cookies and sweets. I did not take the plane to Iraq, but I went by bus from Sinbillawein to Suez or to Nuweiba, then I took a boat. Finally I took a bus to cross Jordan and into Iraq. It took me three days to get there. Arriving in Baghdad for the first time I immediately went to villagers from Kafr al-ʿIshra and stayed with them. They collected ID400, everybody giving what they could afford because, naturally, I did not have money when I arrived. We always do this for the newcomers. They will pay the money back when they have found a job. We also help them to find a job and then eventually they will look for their own place to stay. When I was working in the factory I also slept there. In our leisure time we went to the cinema, we cooked, and I sometimes strolled through the streets in the hope to find a better job.

Iraqis do not fast in Ramadhan. Iraqis also drink whisky and *arak* and beer. Then, they get so drunk that they cannot even find their way home. This is why I did not make use of my driving licence although I have a 'second degree' one permitting me to drive lorries. [He shows it to me.] I did not do that because Iraqis drive even when they are drunk and when they cause an accident the blame is put on the Egyptians.

Iraqis do not work, they are so lazy. After having worked for ten minutes they get tired, they lie down in some corner and sigh: 'I can't work anymore!' [He mimics their behaviour and we burst into laughter.] Many Iraqis open a project and then let the Egyptians work for them. They are so lazy. If they are thirsty they say, 'Give me something to drink!' even if they are standing besides the fridge and I risk burning the cookies in the oven. They would not dare to do this with another Iraqi, they would start an argument. They only do that with Egyptians.

(Amina's husband, Household 4[7])

Many of the points I have made on the foregoing pages explain why the migrants' prospects of returning as made men are not very good. They definitely do not come back with their pockets full of money.[8] Secretly kept, high-flying dreams of a life of luxury and affluence are hardly likely to come true. One reason is that an overwhelming majority went to the two labour-receiving countries where wages are comparably low: Iraq and Jordan. Furthermore, a great majority work as unskilled or, at most, skilled labourers, and many of them do daily wage labour in jobs which are neither very rewarding nor characterized by great stability. Even educated migrants from the village do not find a working place where their education and skills can be used adequately. A further reason is the migrants' short periods of stay, which often do not even exceed a year – a time far too short to compensate for all travelling expenses and to save much money. Not surprisingly, these findings are also reflected in the personal assessments of the villagers:

> It is already some time ago since the *higra* ['migration'] was really rewarding, and when migrants could make lots of money. But today they can only go to these *bilad ta'abana* ['shabby lands'; she is alluding here to Iraq and Jordan]. Today many earn not more than £E2,000 a year in Iraq, or at maximum £E3,000.
>
> (A peasant woman, 35, illiterate)

Another woman said:

> Migration has not really changed the village. This is so because most migrants were away at best two years, and this is too short to make enough money to become really rich. One can marry, but one cannot establish a project. Thus life goes on its normal way.
>
> (A peasant woman, 28, can read and write)

People from Kafr al-'Ishra are realistic about the prospects of outmigration. Yet, in a situation lacking alternatives, migration is still the most promising possibility to augment the household's income. And it is in this sense that migration is important. While people are aware of the outcome of these efforts they still attach their hopes and dreams to them, simply because leaving for work abroad still remains the most promising strategy in a situation in which there is a lack of other income-generating opportunities. Every migrant hopes that he might be luckier than the average, that with only his willingness to work hard he will have a chance to find an honest employer, a well-paying job. It is the 'principle of hope' which, besides the lack of job opportunities at home, drives the migrants to leave. On the other hand, there are of course some migrant households in Kafr al-ʿIshra which have been quite successful. There is one household in the village especially which has turned into the symbol of the migrant household engaged in upward mobility. Villagers over the past years have been able to follow its gradually rising affluence and its material transformation. I will now describe its story.

THE STORY OF THE SUCCESSFUL MIGRANT HOUSEHOLD

The couple is Christian, have been married for nine years, and are in their thirties; they have two boys. Both are civil servants. She works as a primary school teacher; he had an administrative job before he left to work in Austria. His wife renews his leave without pay every year. She is from Zaqaziq, the capital of Sharqiyya governorate, while he originates from a peasant household in the village. He is the only one from his family with a higher education (his two married brothers did not go to school). After marriage they were all living together in the extended household of the senior couple. The house is in the centre of the village and built of mud bricks. This household is exclusively dependent on agriculture on 2 *faddan* of land operated in the *muzaraʿa bin-nuss* system. The wife, whose husband is still in Austria and only comes home for a month in the summer, says:

> When we married I moved into the household of my husband's parents. I had a very, very difficult time and I had many problems. There was very much work when I came back from school in the early afternoon and I was tired and exhausted. Teaching is a very hard job because there are over fifty pupils in one class. Instead of taking a rest I had to start to do all the household chores arising in a seventeen-person household: cooking, baking bread, washing. As I was the only educated daughter-in-law, I did not work in the fields like all the others and so all the household chores were on me. Life was so hard that I ran away to my parent's family several times. Then I talked to the son of my uncle in Zaqaziq. He has been in

> Austria for a very long time and has Austrian citizenship; he has taken his family there. This cousin eventually managed to get a tourist visa for my husband. Now he has been working for six years in the same place together with my cousin. They also stay together in the same flat. In July he will be here for a holiday.

The couple have bought one *qirat* of agricultural land at the periphery of the village and built a three-storey apartment-style house there. The wife had also contributed to the financing as she had sold her bridal gold and the 1 *faddan* of agricultural land she had inherited. Their new house is built of red bricks. The two upper floors are still unfinished and currently used to raise poultry. The migrant's wife continued:

> I immediately moved in when the ground floor was finished. I was very happy because now I could freely determine what to do when coming back from school. I have time to play with my children and to help them with homework. I understand that my sisters-in-law have a very difficult time now because, after coming back from the fields, they have to do what I had done. But a person has to look after himself. And when my parents-in-law knew that I would move out they got very, very angry. But now it is all right. I go to visit them and my husband brings them presents from Austria. We have also bought 1 *faddan* of agricultural land recently. It is his parents who cultivate it and we are doing this *bin-nuss*.

She goes on to explain that *bin-nuss* here means that they divide the harvest on an equal basis, whereas all the inputs have to be borne by the family of her husband. She stores her half of the produce in one of the upper unfinished floors. Compared to most of the village homes hers is quite fancifully furnished, with a crystal chandelier, a huge table, a fan, a colour TV, a double cassette recorder. All the rooms are whitewashed and painted in oil colours, the floor is of concrete and there is a huge carpet. In the glass cabinet there is a complete set of porcelain. She has a modern gas stove in her kitchen and her house is one of the very few attached to the communal water system. To my great surprise I also found a packet of Austrian filter coffee in her kitchen and she was eager to learn from me how to prepare the coffee. She also has a gas oven to bake bread and commented that 'it is much cleaner to bake bread in this oven.

. . . And it is very quickly done; also, I do not need so much help to bake bread.' Her husband has brought a video and next time, she told me, he will bring a video-camera. This video is actually the only one in the village. She has some foreign films her husband has brought from Austria, and some empty cassettes. All the other films she has are of religious content, mostly about the miracles of the famous Coptic saints and the histories of the monasteries. But these are, as she explains, not in her house at present as the priest has taken them with him. She tells me that occasionally she takes her colour TV and the video to the church so that people can watch religious films. The church does not have its own video. I sometimes hear other Christians confirm her story. She also knows how to handle the video.

Plate 5.1 A successful migrant family's house. At the canal we see *gamusa-s*. Some trees have been planted in front of the migrant's house: another sign of status, as is the aerial on the roof. There are other new houses under construction; these are being built in limestone.

In the sleeping room she shows me the dresses her husband has brought for her from Austria and tells me that she puts them on when going to school. She is very proud of them and she likes them.

Then, in a moment, her face gets a serious expression and she asks me: 'Tell me, are these really nice presents or are they just anything cheap and ordinary? Tell me please. You know but I don't.' Her children virtually are the only ones in the village I have seen who possess purchased toys. She shows me two plastic racing cars with remote control. There are many people from Zaqaziq working in Austria she goes on to tell me:

> But here in the village my husband is the only one. When he is here on a visit the sitting room will never be empty of the young men who come to see him. All of them want to know whether he can provide them with a contract and they are very surprised about his luck. We want to finish the two upper floors so that our sons will each have their own apartment and when they marry they can live in them with their wives. It is difficult for me here alone without my husband. But if we only had our salaries from our civil servant jobs that would be nothing at all and we could afford none of all this. My husband says that as long as he is young and as long as he has the chance he has to use it.
>
> (The successful migrant's wife)

On the preceeding pages of this chapter I have described the main characteristics of migrants from Kafr al-ʿIshra. Coming to a conclusion at this point, how can we analyse the characteristics of such a migrant labour force on a globally organized market for labour? In this respect we have to consider again the composition of the migrant community in Kuwait (see Chapter 2). I have pointed out that it is especially the Egyptians, of all Arab migrant groups, who share with Asians those characteristics most searched after on the market – such as relatively short periods of stay, low dependency rates, long working time and low wages. It was therefore no surprise to see that census material from Kuwait indicated that Egyptians constituted the largest Arab national migrant group. In this chapter I have turned to the opposite side of the migratory process by introducing the 'migrant-producing context'. We have thus been able to see that Kafr al-ʿIshra has, in fact, provided a 'pool' of cheap and flexible labour power for the world market for the past one and a half decades. This has been easily deducible from the relative and absolute importance of migrant households in the village. A second salient feature is the high flexibility of this labour power. Many characteristics of Kafr al-ʿIshra's migrants indeed contribute to their resilience: they usually leave only for months, seldom for years; their women and children stay behind in the village; the migrants themselves work as daily labourers and are always ready to change their jobs; they also show a high propensity to migrate again.

Plate 5.2 The village 'car engineers'. Notice that the wheels are made from maize cobs and crown caps of soft drink bottles.

Also important in this respect is the migrant household's integration into agricultural production. As the village economy itself is dominated by agriculture it is perhaps no surpise that the majority of migrant households are in one way or another involved with agricultural production: these households cultivate land and the migrants, prior to their departure, have themselves worked in the fields. In this respect the example of the migrant to Austria is highly significant. This household, with probably the most successful migrant history in the village, was constituted out of a peasant household which disposed of only a relatively small amount of permanently rented land. While household separation and the establishment of a nuclear household consisting of the migrant, his wife and children was the first step to be realized, the second was to buy agricultural land. They are well aware of the fact that after remigration the husband's re-employment in a lowly remunerated government job will not permit them to make great strides and be tantamount to a considerable decline in status. Their future strategy is based on the land they own. It is interesting to note that they have not rented out this plot of land in one of the illegal cash forms which are widespread in the village but in an (equally illegal) form of share-cropping which secures to the household a sufficient amount of subsistence goods. The woman staying behind is, in this case, already practically self-sufficient with regard to the main grain food staples. It can thus be seen that the affluent migrant household also builds its future on agricultural subsistence production. And, even during migration, the household engages in a kind of mixed economic strategy: where cash income from the wife's teaching job and the husband's remittances, and the consumption of subsistence produce from the field (grain) and the household (poultry) are combined. In spite of their relative affluence they did not decide to move to the district town and they did not invest in any form of non-agricultural income-earning project. On the contrary, they built a house in the village with separate flats for each of their boys (who at the time of field research were not even 10 years old), so that after marriage they can live with their families there.

One ultimate 'advantage' of migratory labour power with the indicated characteristics is that whenever it no longer meets a demand on the world market it can principally be reintegrated into the agricultural village and the peasant household. At this moment it is again the subsistence-production-oriented household which resumes the responsibility to reproduce and sustain the migrant's labour power, until perhaps other national boundaries are opened and access to some other labour market is admitted – if not for the migrant himself, then for his son, or perhaps even for his daughter at some time in the future. In this respect it is likewise significant that virtually all remigrants reintegrate into the village and their own households. According to our statistical data only one (0.7 per cent[9]) migrant did not return to the village.

It is the migrant's household which, through its integration into agricultural production, makes the most substantial contribution to the production and reproduction of the migrant labour force. All these households are 'autonomous' in their organization of production and in the allocation of labour power to subsistence or market production. This autonomy also enables them to reintegrate the migrants after their return from work abroad, where they once again can work on the family land and be sustained with the household's subsistence produce. They do not become destitute, uprooted persons who are unable to reintegrate – people who become the responsibility of the state or international charitable agencies.

SMALL PEASANT HOUSEHOLDS AND THEIR STRUGGLE TO SECURE REPRODUCTION: A CASE STUDY OF TWO STREETS IN *HARAT AL-ʿABID*

It is my intent, in what follows, to go beyond statistics. For this purpose I shall introduce the households of two distinct streets in *harat al-ʿabid* ('quarter of the slaves') – the poor district of Kafr al-ʿIshra already introduced in Chapter 4. I shall put special emphasis on the various economic activities each of these households engages in and the relevance migration occupies in this context. These households are typical of the poorer stratum of village society as none of them has a sufficient land base or earns enough money to make them qualify as belonging to the stratum of the middle or big peasantry. Agriculture – as generally in the village – is of overwhelming significance for the majority of the households here. These streets are also typical as most of the households are dependent on the large landowners and operate at least part of their land under the *muzaraʿa bin-nuss* system. They all combine a variety of strategies to secure subsistence, agricultural wage labour being only one among them, practised by most. But there are also households who derive their income from small commercial or manufactural enterprises. Out-migration, as we shall see, also plays a vital role in this respect. All the households are Muslim.

The two lanes are situated parallel to each other and are of relatively recent origin. When the houses were built about fifteen years ago the ground was still agricultural soil at the village boundaries. Lacking building space, young families, in their desire to establish households separate from the parental generation, moved from the interior of the village to this place. Today, these streets in turn are no longer at the geographical periphery of the village, as other people have followed to build their homes beyond here.

Though the houses are relatively new, all are built of the traditional mud-brick material – except for one (10^{10}) which during field research had not been finished. Its future inhabitants, a childless couple in their late twenties, both state employees, live together with the husband's mother in an extended household (9). Deriving their income mainly from government jobs, it is not known when the couple will manage to finish the construction. Houses are one-storeyed except for two. One of them has been painted in yellow and there is the red brick building, all the others are mud-grey. Most houses are built over 1 *qirat* and comprise an entrance hall and two to four rooms. Opposite to the streetside there is usually a small courtyard, perhaps partly covered by a ceiling. Here, stables for the water buffalo, the donkey and the poultry are to be found. There is usually a water pump, an oven and a toilet in this courtyard. Some have jerry cans hanging from the ceiling; these serve as breeding grounds for pigeons. In the majority there is a staircase leading to the roof which is also used for multiple purposes: most important, it is a storage place for straw, used as firewood and dry animal fodder; women dry their laundry and may also store home-produced food items on it. Some of the houses have a small front-door veranda where people spend their evenings, where women do their household chores, and where pupils write their homework. Houses are built adjacent to each other without free space between them. There is one empty site (2) which the owners (8) use in shady hours to tie up the water buffalo or where agricultural produce is stored. There were two empty houses at the time of field research. One (13) belongs to a family which has left the village. The other (14), comprising a very small house and a courtyard, is owned by a household (3) which plans to build a larger house and move in here in the future. Only one other household (16) has a relatively large open space. Streets are usually filled with life: toddlers crawling in the dust, children playing with their home-made toys, women standing together for a chat or doing their housework, men smoking their water-pipes. But I have also strolled along these lanes and found them virtually deserted with all doors and windows closed – this is the time of heavy agricultural work when everybody is out in the fields.

Let us first describe some important demographic characteristics of the households in these two streets. A total of sixteen families live there, of which five[11] are extended. These extended families are headed by an elderly couple and comprise young sons and daughters, as well as married sons and their wives and children. The couples of the nuclear families are all in their late thirties. They have between

Map 5.1 Two streeets in *harat al-ʿabid*

16 ext fm	17 Sausan nucl nm	18 Hagg Salim and Aisha rm	19 Nadia nucl rm	20 Fauziyya nucl rm

11 ext rm	12 Farida nucl m	13 Empty house	14 Empty house	15 Mustafa nucl nm
6 nucl nm	7 Sahar nucl fm rm	8 Maher and Fatima nucl m	9 Im Rida ext rm	10 Unfinished red brick house

1 Ibrahim ext fm, rm	2 free space	3 Rida nucl m	4 Amina nucl m	5 Majid and Muhammad ext nm

Key: nucl nuclear household; ext extended household; fm future migrant; m migrant; nm non-migrant; rm remigrant two and six children. The high rate of infant mortality here is striking: out of the sixteen households there are eleven who reported that at least one of their children had died, and there are only three which stressed that all of their children are alive.[12] Some families are related by kinship ties. A widow (9) has two of her married sons living opposite to her (3,4), while her third son and his wife, owners of the unfinished red brick house (10), stay together with her and a divorced sister. Household (6) comprises the father of household head of (7); this man is also a brother of the household head of (11), who in turn is father of the head of household (12). The female head of household of (17) is a sister of household head (20). Seven[13] of the households are headed by women: due to divorce (17) or because of the death (5,9) of the former household head; the other four households (3,4,8,12) are temporarily headed by a woman as a result of the migration of the family father. Four other households (7,18–20) have been formerly headed by women due to their husbands' migration. Thus eleven out of sixteen households (68.75 per cent) are permanently or temporarily headed by a woman.

Illiteracy or near illiteracy is still rampant in the two lanes. There is no educated head of household here. The only persons who could figure as such in the near future are the couple to whom the red-brick house (10) belongs, currently living in (9). He is the only educated child of his mother and his wife comes from an urban area. In an extended household (1) the only educated son has also married an educated woman, she is from Mansura. These four persons are the ones with the highest formal education as they all hold a diploma.[14] Most of the male household heads did not go to school, and if so for at maximum three to four years. Only one woman of this generation (15) completed primary. Women are illiterate, as are in fact most of the men, even if they went to school. But there are also men and women who manage to read and write fairly well without having had any formal education (e.g., 18,19). Not all of the children under the age of ten go to school (3,4,20) and most girls drop out very soon after, at maximum completing primary school. While there are some pupils who might in the future go beyond this and reach high school I got the impression that people in these streets are generally not very interested in the education of their children.

Agriculture

The main basis of a secure future for these households is agriculture, as twelve[15] out of sixteen (75 per cent) derive a substantial part of their income and subsistence produce from work in the fields, be it on their own fields or/and as agricultural wage labourers. There is only one household (5) which during field research had no connection with agriculture, though its occupants raise poultry for home consumption. Ten households have permanent access to land.[16] Out of the six landless households,[17] there is one (19)[17] which occasionally rents in some *qirat* for a season, usually to grow clover or maize to feed its water buffalo, or rice and wheat for home consumption. Acreages range from a mere 8 *qirat* to 3.5 *faddan* with an average of slightly above 2 *faddan*. Three[18] of these landless households derive their income predominantly from agricultural wage labour. Members of all households – with the exception of (5,15) – do agricultural wage labour occasionally. Small children work in the cotton worm campaign or in picking tomatoes, women work in the harvesting of wheat and maize, men bring in the rice, to give only some examples. If there is labour in the fields anybody who is free, not too young or old or too ill, will seize the opportunity and earn some additional income.

One household (11) had a camel which transports crops from the fields. The owner, guiding his camel on the ways to and from the fields, is not remunerated in cash but in kind. Transportation of wheat straw of 1 *faddan* to the cultivator's house is rewarded with eight *kila*.[19] The same household also has about forty sheep. Nine out of the sixteen households have a buffalo or at least a calf. One of these buffaloes (20) is maintained in joint ownership with a wife's brother. All households engage in some forms of home-subsistence production, they raise poultry, conserve vegetables and dairy products.

Off-farm activities

Four households[20] derive their income mainly from a small enterprise. Mustafa (15) has established a workshop where he repairs rubber tubes of the diesel-fuelled water pumps which are now widespread in agriculture. As the tubes seem to be highly susceptible to damage, and as there are no similar workshops in the surrounding villages, the project is prosperous enough to employ three workers. Though almost permanently present, these are remunerated on a daily basis with £E3–5. The household also has the income from 2 *faddan* of agricultural soil, which Mustafa's wife inherited from her father. She has rented out this land under different (illegal) conditions. Part of it is cash-rental on an annual basis, the other part is rented out on a share-cropping, seasonal basis. The income of the second *bin-nuss*[21] makes her virtually independent from the market, as she derives from it a sufficient amount of wheat and rice, the two most basic food staples. Thus this household, whether from the workshop or from rent, also derives its income from agriculture-related production and services. The household is one of the more prosperous in the two streets under study. As in all the other families, poultry is raised.

There is also a tailoring workshop (5), run by Majid and Muhammad, two brothers. This household is the only one which is in no way engaged in agricultural activities, its occupants' livelihoods being secured entirely from the income of the tailoring of men's *gallabeyya-s*. The family has separated one room from the rest of the house to serve as a tailoring workshop; it has a separate entrance so that customers can visit the enterprise without having to interfere with family life. This shop was established by the late head of the household and is now run by his two married sons, both in their twenties. During the month of Ramadhan Majid takes his sewing machine and assets and goes to work in a village in the governorate of Buhaira where he settles down in a relative's home to receive customers. As there is no tailor, and as there is usually a lot of work during this month, he sees it as a chance for an additional income. The brothers hand over all their earnings to their mother, the head of the household, except for a small amount of pocket money, mainly used for cigarettes, hashish and their children's sweets.

Though it will take me away from my general theme of non-agricultural activities I cannot but point to the strange difference observable in the self-representation of the male and female members of this household. It is here that the interplay of 'modern' and 'traditional' life-styles becomes most visually apparent. While its female members are all dressed and move about like most of the village women the sons most obviously lean towards 'modernity'. Both sons often wear coloured track suits of synthetic fibre, they also wear fashionable, urban-style *gallabeyya-s* – usually white, form-fitting, sometimes short sleeved – the kind Rugh (1989) has called the '*frangi* or *scandarani* sophisticated style'. Their little girls are sometimes dressed in an equally 'foreign' manner in lace dresses, and a 2-year-old boy can be seen in jeans and shirt. As skilful tailors the two brothers quickly take up new urban fashions and materials and introduce them into their work. While their wives are practically invisible, always very busy with housework and usually not looking very happy, their husbands, strolling through the village and trying to talk to some young girl, often gave me the impression of young unmarried men rather than of family fathers.

As there is no public café in the village, the workshop has come to serve as one of the meeting places for men. Endless rounds of tea are invisibly prepared by the daughters-in-law and served by the two brothers. While listening to Egyptian pop songs from a cassette recorder, their visitors exchange the latest news from the village. The workshop thus also functions as a market for information, and here, migration is an important issue. Though they themselves have not been working abroad the brothers have profited considerably from migration: as one of the favourite presents of return migrants is textile material, villagers come to have it tailored by the two brothers.

Sausan (household 17) also derives her income from a non-agricultural enterprise. Divorced by her husband some ten years ago, she engaged in vegetable trading in order to provide a living for herself, for her old mother, and for her three children. It is not surprising, that Sausan, lacking agricultural assets, turned to petty trade, a typical economic strategy of women. After she and her husband became separated she initially found shelter in the house of a brother, but eventually managed to save enough money to build her own house from her savings. Only recently she added a room on top of this house, to prepare for the marriage of her only son. Over the years she has been quite successful and at the time of field research she ran one of the largest vegetable petty trade enterprises – which means that she sometimes has about 100 kilos of merchandise on her veranda. Her trading enterprise not only provides her with cash – a considerable number of customers pay in kind and she then decides whether to consume this produce herself or to sell it again at a later time. Yet, in spite of her relative success as a trader agricultural wage labour is still more important. In peak seasons she usually closes down her enterprise and works in the field which earns her more money than trade, she says. All her children have

Plate 5.3 The tailors' workshop. In it are signifiers of a modern life-style: both men wear watches, there is a cassette recorder, a packet of filter cigarettes, and photos of Egyptian pop singers. But there is also, to the left, a small photo showing the Virgin and Child, as well as the name of Allah and Islamic verses.

to help her and they also occasionally work as day labourers in agriculture. While the tailor's workshop, as we have seen, functions as a café for men, her veranda is a meeting place for women who not only come to buy from her but also come round for a chat. Unlike the men, the women are always busy with something, perhaps cleaning *ruz* (rice) or *ʿads* (lentils) from little stones or dirt.

Farida, living opposite to her in a landless household (12), has engaged in two different economic activities. A woman in the village taught her how to sew women's and children's clothes. She went to this woman's house for a year where she did the household chores while at the same time acquiring sewing skill; she also has a little *butik* where she sells foodstuffs and household items to the villagers. One person (9) has a stable monetary income from a government job; the young couple (1), have both acquired diplomas and are waiting for their appointments to jobs as civil servants. Three men occasionally find jobs as construction workers or in a factory (1,18,19). One elderly woman (1), every second day, helps in the household of George, one of the richest landowning families in the village. She is remunerated in cash and kind.

It emerges very clearly that agriculture still figures predominantly as a source of income in these two streets of *harat al-ʿabid*. Returns from work on the holding are nearly always supplemented by daily labour in the fields of others. But there are also some small off-farm enterprises, three of them belonging to the more traditional activities in the village, while one is linked to the recent mechanization of agriculture. The preceding compilation of agricultural and non-agricultural economic activities pursued in these two streets might perhaps appear confusing. Yet this situation actually mirrors the practices of these households which, in ever-changing arrangements, combine different economic activities: a plot of land is rented for one season and dropped in the next, occasionally a plot of one's own land might even be rented out, a water buffalo is bought or sold, a child is sent to school today and to the fields tomorrow. It is exactly this diversity, and constantly changing combination of different income-generating activities, which characterizes these small peasant households' economic flexibility.

OUT-MIGRATION FROM THE TWO STREETS IN *HARAT AL-ʿABID*

If agriculture is of overwhelming importance in the two streets, so is out-migration. At the time of field research, twelve households out of sixteen were directly involved in migration, be it that they had a member who previously worked in an Arab country, who currently was abroad, or who had concrete plans to migrate in the near future. Out of these twelve households: seven (1,7,9,11,18–20) had a remigrant; four (3,4,8,12) had a member working abroad; and in three (1,7,16) a member was about to leave.

Four households (5,6,15,17) had no migrant nor did they have concrete plans to send a member abroad. Thus the percentage of migrant households (i.e. comprising a remigrant, a person about to leave or a person absent at the time of field research) is 75 per cent, compared to 43.2 per cent for the entire village. Five remigrants[22] had been exclusively to Iraq. two exclusively to Jordan (8,12), one (19) had been only to Saudi Arabia, and one (18) had been to Saudi Arabia, Iraq and Jordan. Most of them had migrated more than once, but not more often than three times. It is noteworthy that there are also two cases of internal migration. One (19) went with his wife to Cairo where they stayed for one year and he worked in construction. They eventually returned to the village because they found conditions of life too hard in Cairo and wages not worth while. It was only after their resettlement in the village that the husband went to Saudi Arabia. The second case (9) is of special interest as here the wife also migrated for work. This was before their marriage when she worked in a bread factory in Cairo for two years. During this time she stayed in the apartment of her brother. As I have previously stated that village women do not migrate it has to be noted that she is not from Kafr al-ʿIshra but from an urban area. While she worked in Cairo her future husband was in Iraq.

The figures suggest that in the two streets migration seems to be the norm rather than the exception. It is therefore interesting to investigate into the reasons which might cause the remaining households to refrain from work in a foreign country. Let us have a closer look at these four non-migrant households in the two streets. One is the household of Sausan (17), the vegetable seller. Her eldest child, a 17-year-old youth, in fact now and then speaks of work abroad. Asked for her opinion about the future migration of her son the mother says:

> I won't let him leave now. He is still too young. And I need him here because he is the only male in the house. I have now built a first floor on our house because he will soon marry, and then he will stay here with his wife.

We have previously seen that women in Kafr al-ʿIshra generally do not migrate. The household therefore has no one who can work abroad and is tied to the village economy in its struggle to secure subsistence. For the time being her son does not express any concrete interest in migration due to his youth which does not yet qualify him as a 'typical' migrant.[23] What perhaps is of greater importance here is his mother's interest in keeping him at home. If her son was no longer with her the household would have lost the only male and a considerable potential of labour power. While she would probably manage without him her success is also definitely dependent on the relatively large pool of labour power provided by herself and the three children. The internal division of labour within her household assigns large parts of the housework to her two daughters which sets her free to concentrate on her trade activities. She often goes twice a day to the commercial centre of the district town of Sinbillawein to fetch goods from a wholesale trader – which is a rather time-consuming job. Another reason for her relative success is that not only does she offer her vegetables in front of her own house but she often also sells her goods in the centre of the village, thus catering to the needs of village women who are not inclined to walk as far as her home, or – and this might also be a reason – wish to avoid setting foot into *harat al-ʿabid* as this is 'enemy territory'.[24] It is here that she needs her son. While one of them is sitting at home to serve the customers and to look after the goods the other one sells in the village centre. She also lets her son fetch the goods she has bought in town. If there is work to be found, the children are also hired out as daily wage labourers. Thus all of her children contribute monetary income to the household.

Her strategy as household head is based on the enlargement, or at least conservation, of the available amount of labour power. Any reduction would be concomitant to a loss of earnings. Her strategy is also visible in her plans to marry off her son, which will provide her with the additional labour power of a daughter-in-law, especially as the eldest of her daughters is nearing marriage age. It may then be the household head's desire to secure the maximum amount of labour power, as well as the son's indecision to migrate, which will make the household figure among the non-migrant ones.

Another non-migrant household is that of Mustafa who owns the repair workshop (15). As the only adult male he is responsible for the running of his enterprise. He does not want to migrate:

> No, I do not want to go to Iraq or Jordan. It is better to stay at home than in the *ghurba* ('to be in foreign parts'). I have my work here in the village which feeds me and my family. I plan to buy a taxi soon, and then I will leave the work to my labourers while I drive my taxi. Who guarantees me a job in Iraq or Jordan? Here I have everything: my work, my family and my parents.

In this household the only male member runs the enterprise. He definitely sees his fortune in the village context and not in migration to Iraq or Jordan. As his workshop is prosperous enough to allow him to plan investment into a third source of income besides landed property and the workshop, migration plans do not figure in his immediate future. In his case income-earning prospects abroad seem to be far more risky and less rewarding than in the village. Also, he is tied to his workshop as there is no other person who might take his place. He could not divert his labour power, experience, connections, and authority from the workshop without endangering its management. Migration to him would mean to abandon a 'lucrative' and secure enterprise which still seems to be in the process of expansion.

Another non-migrant household owns the tailor enterprise (5). Though less prosperous than that of the repair shop of Mustafa, similar arguments could be forwarded here. It is also noteworthy that Majid occasionally practises a form of internal migration. The last non-migrant case (6) is a recently established nuclear household. The household head is already in his forties and had been living with his first wife, and later with his second wife and children, in the extended household of his father. When the latter died, he and his brothers and their families stayed together in the house and farmed the land jointly. When he finally separated from the parental household he started to cultivate his share in the heritage on his own. It comprises 2 *faddan*, operated in the *bin-nuss* system, as well as 1.5 *faddan* of owned land. Asked whether he plans to work abroad he said:

> I would love to work abroad, like all the young men do nowadays. But now I am too old. I should have gone when I was still young. My children need me, I can't leave their mother alone with them, because they never would listen to her. They need a strong hand. And who would care for the land? Praise to God, we have everthing we need.

One major impediment to migration is age – thus reflecting the 'market laws' and the interest in young labour power only. Only 6.9 per cent of all migrants from Kafr al-ʿIshra were above the age of 40 at the time of their first migration. It is in this context that we can interpret this man's statement.

It is very noticeable that in the two streets examined here the pool of prospective migrants who had not been abroad at the time of field research is virtually exhausted. Every household with a male of working age considers and discusses out-migration, and in the majority of cases migration occurs. If this is not the case, then plausible reasons are discernible. The most important impediment for migration, needless to say, is the household's lack of a person qualifying for migration in terms of sex, age, but also for health reasons. But there are also those households which have chosen a village-based income over migration. The examples of the tube-repair workshop and the tailors show that households weigh up the different possibilities. Migration today is not coercive and there is no 'natural law' whereby households displaying specific socio-economic characteristics inevitably set free a certain amount of their labour force for migration. Households see in migration one strategy among others and they actively assess it and its chances for success in comparison to others. If a household possesses assets, be these agricultural land, animals or a small enterprise which secure a stable and safe income but which could not be maintained without risk in the case of the withdrawal of labour force, then the household is likely to opt against migration – at least under the existing prospects for successful work abroad. Here, deciding against migration, especially in a situation of declining income-earning prospects abroad, appears the safer strategy. This is also clear from the following experience:

> I live together with my wife and my children. In the past I stayed with my brothers and sisters and my mother. Naturally, as it is the tradition here, we separated the household and divided the agricultural soil among us. I got about 3.5 *qirat* of building land and 16 *qirat* of agricultural soil. I built a house for my family, consisting of two rooms and a hall. I am a bee-keeper, and I work for people who have beehives. At the end of the year I get a quarter of the honey. I also cultivate my land. I save the money I earn from the sales of honey. I made the bricks for my house on my own. Eventually I bought some additional agricultural land, and today I have nearly 2 *faddan*. I also bought a beehive for myself. We have a cow. We consume half of the milk and the milk products, and we sell the other half. . . . My wife raises poultry in the house and we eat a large part of it, the rest we sell on the market. I never thought to travel: I was unable to go to a foreign country because I could not abandon my land and my bees. Had I

gone abroad, I would have had to sell everything. After my return I would not have had the chance to buy all that again because everything is getting more expensive. I also do not want to leave my children and my wife alone. Many agricultural labourers have gone abroad. These people would do better to establish a small project here in order not to abandon their land.

(Peasant and bee-keeper, 41, can read and write)

CONCLUSIONS

In this chapter I have pointed to the ways in which Kafr al-ʿIshra's migrants meet the interests of a globally organized labour market. Shifting my focus to the micro-perspective I have investigated the relevance of out-migration as one economic strategy among a variety of others households engage in to secure their survival. It has become clear that in the 1980s agricultural production and labour migration have, in fact, been the most important economic activities of nearly all the households I have been dealing with here.

Taking the households of two streets in *harat al-ʿabid* I have shown how they function as places where commodities – be these cash-crop or migrant labour – and subsistence goods are simultaneously produced. It is this combination of different forms of production and income-generating activities which enables these poor households to survive in very strained economic conditions, and also allows them to reintegrate the remigrants after their return from abroad. It is these functions we have to concentrate on, rather than seeking the predominant causes in an imbalance between push and pull poles which result in an over supply, if we want to understand why Egypt's peasant migrants may be bought cheaply on the market and flexibly deployed. Naturally migration changed the householders' lives in many respects for nearly all the households in the two streets which had a member with the major migrant characteristics (such as age and sex) in fact sent this member abroad for work. Yet withdrawing a considerable amount of labour power from a household's production and income-earning activities does not take place automatically but as a result of a decision-making process. Thus, the existence of a promising, safe and stable basis of permanent income in cash or kind will allow a household to abstain from setting free part of its labour force for migration. In this case the fear that the absence of part of the household's labour power could seriously threaten to damage or destroy this major source of income leads to a negative migration decision. This has given us an idea about the autonomy with which these households organize their economic activities.

It was Meillassoux (1976: 112; see also Wallerstein 1984) who convincingly argued that migratory labour force is cheap because the production, and the valuation, of this labour power takes place in different economic sectors with different forms of production prevailing. Applied to the Egyptian case this means that its peasant labour power can be bought for a cheap price, not because virtually every adult male wants to migrate or because of a resulting oversupply but rather because this labour power is inexpensive as it is produced by the subsistence-oriented peasant household and hence its production costs need only partially be borne by the capitalist employer in the Gulf states. In the intention to reach a more profound understanding of the 'functioning' of migration, we have therefore proceeded to the level of production, instead of concentrating on the level of circulation (see Meillassoux 1976: 112). The place of production is the village community and it is the households which have the capacity to produce and to maintain the labour force. It is its production and maintenance by the peasant household which makes this migrant labour power so attractive for capital. And it is here also that we find the causes which make labour power move spatially. Meillassoux highlighted the fact that the peasantry's labour power may be particularly cheap if it is produced in a non-capitalist form of production. It is inexpensive if its production and maintenance take place within a household which is still organized as a unit of production for consumption. These households, even in a time of advanced global market integration, are only partly integrated into the monetary sphere as household production for consumption is not remunerated in terms of a cash wage. This labour, therefore, remains without 'value' (Elwert and Wong 1981: 226).

It is clear that this is a special characteristic of the peasant household which has at its disposition a sufficient input of means of production in terms of land, tools, animals and labour force. Be it for labour migration or for cash-crop production, as for any other form of wage labour, the labour force from such households is so profitable within the capitalist mode of production because it is exploitable in a double sense, as has been pointed out: in the 'conventional' way within capitalist production where surplus value is generated from wage labour (Meillassoux 1976: ch. II, 5). Applied to our case this, firstly, refers to any kind of wage labour that household members do in the village as well as to the wage labour migrant-household members perform abroad. This generally is the case within the capitalist form of production where labourers are deprived of their surplus value. Secondly, and this is the characteristic feature of the *ʿizba* system of production, capitalist employers need not cover the costs of the (re-)production of the labour force as these are externalized to the subsistence-producing peasant household. Subsistence production, in this sense, has to be understood as production for direct consumption, hence production of use values:

> In *all* societies (not only in the 'primitive' ones) there is the realm of production for survival, discernible from other realms of societal activities: *subsistence production*: we do *not* mean a specific, historical form of production, but generally all productive activities . . . which are linked to the preparation and processing of household inputs (household understood as unit of production) in the direct sense, that the materialized labour is not sold outside of the unit of the household. It is in a direct sense necessary labour: labour which directly serves the keeping-alive of the household members across the generations.
>
> (Schiel and Stauth 1981: 127; my translation)

The capitalist employer can therefore also pocket (at least part of) the workers' labour rent.[25] Here, then, lies the principal reason for the cheap price of the migrants' labour. International labour migration could not be as profitable without the possibility of an externalization of the costs of production and maintenance of the labour force.

Central to the analysis, then, is the persistence of the household as a unit of use-value production within the general, global capitalist mode of production and the role migration plays in this context. There has been a long scholarly debate about the fate of the peasantry's subsistence production within this general framework which need not be elaborated on here.[26] Older theories claimed this sector's final dissolution in favour of a capitalist form of production (Lenin 1967).

Very similar to the arguments put forward with regard to the general impact of the capitalist market integration on subsistence-producing milleux, it has often been argued that out-migration destroys the vitality of these households. According to this perception, migration, as one aspect of market integration, leads to the peasant household's eventually losing control over organization and means of production, a process which is mainly linked to two causes. One is the siphoning off of its labour force into wage labour (or any other form of labour, such as forced labour or military service), resulting in a 'strategic deficit' (Evers and Schiel 1981: 293) – a deficit in manpower which affects the household's standard of living and, indeed, survival. This unavailability of a sufficient amount of labour power leads to a second major threat to subsistence production: the need for cash. As the household is no longer able to carry out the necessary working tasks it has to turn to the market to buy the required goods. The general need for an increasing amount of cash, however, is not restricted to the aforementioned case. Other reasons are, for example, the developing of new 'modern' consumption needs, the monetization of social relationships (here the steadily increasing bride-price of Muslim marriages is a case in point), or increasing taxes (Elwert and Wong 1981: 260). As migration is just one other consequence of market integration, similar effects are to be expected.

In this sense Meillassoux (1976) has referred to the eventual decline of the 'household sector' as a consequence of migration. He argued that the (remaining) members' struggle to continue the household's production during the migrant's absence leads to the twofold exhaustion of the producers and the soil as both slowly lose their potential to secure a living.

> Therefore . . . a growing amount of the migrants' remittances is used for the purchase of foodstuffs on the market or, in other cases, these are used for the employment of seasonal labourers for the cultivation of fallow fields.
>
> (Meillassoux 1976: 147; my translation)

As the migrant labour is determined by instability and job insecurity the result, according to Meillassoux, is predetermined:

> When the monetary income of the migrants no longer compensates for the falling off in agricultural production then misery gains ground until finally conditions of physical reproduction of the workers are endangered.
>
> (Meillassoux 1976: 150; my translation)

Contrary to these approaches the functionality of the peasant household's preservation as a unit of production within the global capitalist market has to be stressed (Evers 1987, Smith *et al.* 1984). In fact, today it is known that without the existence of a 'subsistence sector' capitalism would not be possible at all. This is quite evident even with regard to the functions households have in the centres of capitalism – though these have been almost completely stripped of their means of production. The process, however, cannot be completed. This is the case even in those places where wage labour has become the predominant form of production. Here women's functions within the household's economy have to a large extent been reduced to the refinement and processing of foodstuffs purchased on the market. But even here, women cannot be stripped of their functions as (re)producers of life: they bear the children. These processes, of course, have not (yet) been of equal significance in large parts of the Third World. It is here that the agricultural sector still pursues its 'traditional' tasks to a large extent. Hence the maintenance of subsistence production, an economic sphere of unremunerated labour, plays a vital role within the new international division of labour.[27] Having had a close look at the households of these two streets it is clear that virtually all of them are involved in remunerated labour while at the same time producing subsistence crops which enter the consumption cycle of the household.

Another salient issue is that the logic of the system of contemporary international migration is based in its balance – that is, in the upkeep of a *status quo* where the peasant household functions as a unit of production of a 'free' (i.e. unremunerated) labour force which is available principally for any sector it is needed in.[28] The functionality of the subsistence sector is thus dependent on a kind of 'stability of instability'. It is a matter of fact that in many Third World countries, and definitely in the Egyptian countryside, we can observe a constant trend to weaken the household's basis of reproduction without however allowing for its final disintegration. It is indeed the continuity of the crisis which is an important inherent characteristic of the system. In this sense Evers and Schiel argue:

> An elimination of these impediments for full-scale valuation of this work force, which would be tantamount to the total 'modernization' of the traditional structures, is a priori excluded for easily discernible reasons: this would be connected to the

> destruction of the basis for the 'valueless' reproduction of labour force, on which the specific production of surplus value of the market-oriented sector (i.e. also the reproduction of the specific relations of production) is grounded.
>
> Therefore often a 'conserving' restructuring is to be observed parallel to the partial dissolution of the old societal structures. The result of both processes is the transformation of the society towards a seeming 're-traditionalization'. Instead of 'modernization' 'traditional' institutions emerge, which at the same time delay the total disintegration of the society organized on subsistence production and also 'help' to perpetuate the strategic deficit.
>
> (Evers and Schiel 1981: 293–4; my translation)

The subsistence-producing household, according to this logic, has to be kept in a precarious balance between two poles, one of which would be autonomy from the market. If these peasant households could reach an economically independent status, enabling them to secure their continuity basically from the consumption of their own produce, they would be no longer in need of major additional monetary inputs from wage labour. They would come near to being autonomous in their decision as to whether, to what degree, and in which form, to integrate themselves into the wage labour market; as a result their functionality for the capitalist sector would thus become potentially endangered. Though this is only an ideal type which is not really likely to be realized, any danger of such a development has to be excluded. In this sense Luxemburg (1969) had already stressed that the destruction of the independent character of what she called 'societies of natural economy' is inherent to the process. She was perhaps the first to recognize the importance of the 'non-capitalist' production when she said that 'Capitalism needs for its existence and development non-capitalist forms of production in its surrounding' (Luxemburg 1969: 340), and that

> Without their [i.e., formations of 'natural economy'] means of production and their work force it [i.e., capitalism] cannot exist, nor without the demand for its surplus product. But in order to extract their means of production and their work force, to change them into buyers of commodities, it systematically strives to destroy them as independent social formations.
>
> (Luxemburg 1969: 343)

Mitchell (1991), in a recent article addressing the case of the Egyptian peasantry, has shown in which ways political agency, especially the state and international development agencies, consolidates the perpetuation of this 'stability of instability' by creating the necessary legal, institutional and aid frameworks. Rather than aiming at autonomous self-sufficient reproduction the constant siphoning off of resources and values to the profit of outside forces is secured.

Turning to the other pole of this precarious balance the second possible 'danger' for the peasant household's function within the capitalist system lies in its full proletarianization (i.e., the total abandonment of subsistence production). In this case, labour rent would no longer be extractable from the workers since the price of their labour must then include the costs of reproduction of their labour force as well as that of their households in general. An important characteristic of these households and their 'floating' within a situation of 'stability of instability' is their great flexibility and resilience – here with reference not to their deployment on international markets but to the organization of subsistence production. These household are not, as one is inclined to deduce from the writings of Meillassoux, static and unchangeable in their 'domestic mode of production', knowing no other reaction than to allow for their weakening in the face of the onslaught of the capitalist mode of production.[29] Subsistence-producing households are not 'antiquated' (Meillassoux 1976: 106) as he has characterized them, as I have shown in the empirical examples presented in this chapter. The 'stability of instability', then, refers to a state of deficiency which necessitates a virtuoso combination of an everchanging variety of strategies, and here the constant, but partial, integration into the wage labour market, whether in the village, its surroundings or abroad, plays an important role. It is within this context that we can understood the economic practices of subsistence-producing small peasant households in *harat al-ʿabid* which I introduced in this chapter. In this sense Schiel and Stauth (1981: 134–5) have argued that it is indeed the 'traditional sector' which is changing most rapidly in a modernizing sense, if one was to apply the criteria of modernization theory. It is the subsistence-producing sector which is the most innovative in order to secure its survival. In this sense Wood (1982: 312) conceptualized the 'dynamic character' of the household.

To conclude, it has been very evident that only the very lucky find a wage labour job guaranteeing a sufficient and stable income which allows a household to drop out of subsistence production. Certainly no household in the two streets was able to do so and all of them still struggle for survival by combining different and varying forms of production to make ends meet. This, then, is the case in present-day Kafr al-ʿIshra in general. It is not a 'traditional' village in the sense that it hasn't joined the general trend towards capitalism and because of its inherent forms of economic and cultural production. Kafr al-ʿIshra is not a case of a community which has somehow been forgotten in an overall trend towards 'modernity' and it is not 'backward' due to some hidden characteristics of its inhabitants. Kafr al-ʿIshra, its people, and its forms of production, are part of the modern world, of the globally organized mode of production. Within this context, the production of use values is seemingly only a 'traditional' and surviving form of production, rather it is a constituent aspect of the overall capitalist system.

Out-migration from Kafr al-ʿIshra shows that the *ʿizba* system of production has been spatially enlarged. Until the beginnings of the oil boom in the first half of the 1970s the *ʿizba* system of production had more or less been restricted to the village site, where cultivators lived and simultaneously produced subsistence- and cash-crops. After this turning point the place of deployment of their labour force has also been situated further afield. Today, the subsistence-producing peasant household's labour force is not only deployed on plantations and fields where cash-crops are produced, but simultaneously works on sites in foreign countries. Kafr al-ʿIshra is an example of the increasingly worldwide extension of the *ʿizba* system of production.

6

MIGRATION AS A STRATEGY OF THE SMALL PEASANT HOUSEHOLD

One of the main concerns of the last chapter was to investigate the factors which made migratory labour attractive on the world market. Both the flexibility and cheapness of migrant labour have been grounded in the peasant household's articulation of different forms of production: members of these households perform various kinds of unremunerated labour at home and in the fields, their members produce subsistence and cash crops, they work as daily labourers in the fields, in the district town and abroad. I have also pointed to the fact that the perpetuation of this system is linked to a kind of 'stability of instability', which hinders peasant households from becoming entirely wage-labour oriented or completely self-sufficient from subsistence production. I have already discussed in the conclusion of the last chapter the fate of the subsistence-producing context within market integration and I have argued that rather than being finally dissolved these households will survive as a result of the consequences of labour migration.

Generally speaking, monetization, through partial integration into the wage labour market, cites as one instance for the dissolution of 'traditional' forms of production, has long since been a characteristic of the subsistence-producing household. And the postulated gradual externalization of the household's labour force leading to its eventual *Entdinglichung*, its loss of substance, is actually only one possible outcome. The other possibility is that wage income may be used for the reinvigoration of subsistence production.[1] In this context, labour migration becomes of great importance. This, then, leads us to the major questions of this chapter. Provided that part of the peasant household's labour force migrates voluntarily, what then are these households' goals and aims, which strategies do they follow? And, inseparably linked to this question, what do these households do with the money earned abroad?

The notion of migration as a household strategy immediately calls for some clarification. Until this point the concept of the household has been implicitly based on the notion of a quasi-monolithic unit. Here it seems to be likewise taken for granted that such units have one unique goal, one unique migration strategy. Yet households are composed of different members and their ranking is not equal. Without doubt the household head's position is endowed with authority and power to an incomparably greater degree than that of, say, an unmarried daughter. Viewed from this angle it becomes questionable whether such households are appropriately defined as homogeneous entities. And, given a hierarchical order and a distribution of roles within the household, the question also comes up whether it is appropriate to depart from a uniformity of interests. It has therefore to be clarified in which sense one can speak of homogeneous household migration strategies or whether there exists a diversity of goals and strategies pursued by members occupying different statuses and roles within the household.

INVESTMENT INTO AGRICULTURE

The interrelationship between agriculture and the migrant household has been a core issue of the last chapter. There, I have shown that a majority of the migrant households is in one way or other integrated into agricultural production. Now, the question comes up to what degree migrant households' goals and strategies are also related to agriculture. Is the investment of remittances into agriculture a strategy of these households? Or do these households struggle to realize a non-agricultural economic basis? The case of the 'successful' migrant in the previous chapter has already indicated that the consolidation of subsistence production seemed to be a major aim. To develop my argument I shall present in some detail the cases of two remigrant households in the two streets introduced in the preceding chapter (see Map. 5.1, p. 121).

The case of Maher and Fatima

Maher and Fatima settled in Kafr al-ʿIshra a couple of years ago. They had been previously living in the extended household of Maher's father in an adjacent village. One reason behind this decison to settle in the wife's village was a plot of land they cultivate. These 16 *qirat* belong to Fatima's father and are operated in the *muzara'a bin-nuss* form. Maher has been working in Jordan for five years. They first invested remittances into building a house of their own in Kafr al-'Ishra (8), and also bought the courtyard (2) opposite to their house, a place where in the future a home for their children and their wives can be built. It now serves as a place for the *gamusa*, the water buffalo, and its calf, as well as a place where agricultural labour, such as threshing or bundling of straw can be done. The *gamusa* and the calf have also been bought from remittances. They have also invested into half a *faddan* of agricultural land and have bought a water pump which they rent out to villagers. He visits his wife once a year for a holiday; in the meantime Fatima replaces her husband as a head of household. She not only works on their own fields and cares for the cattle but also engages in additional agricultural wage labour. She also helps her neighbour if the need arises.

Maher and Fatima obviously invested a substantial part of their remittances into agriculture. Their migration strategy has been moderately successful, and while far less well off than the civil servant couple in the last chapter they still represent the more prosperous and promising of the migrant households.

The case of Hagg Salim and Aisha (18)

Hagg Salim is 36 years old. He is married and has four children. He is from a poor, landless family and his father was a small vegetable seller who turned the villages around Sinbillawein to sell his goods. When Salim's father died, his mother soon married another man and went to live with his family. As the latter did not accept her earlier children, as is often the case, she had to leave them in the care of their eldest brother. The siblings neither had soil nor animals and in order to provide for a living they were totally dependent on the scarce opportunities of wage labour in the village. Salim never went to school, but he managed to acquire some basic literacy on his own. He got married early and his wife joined him to live with his brothers and sisters. To escape from his *azma* ('crisis'), as he calls it, he did not see any other possibility than to migrate. Salim was lucky to procure for himself a contract through the mediation of a fellow villager. He thus went to Saudi Arabia and worked there for two years. Later he went to Iraq and to

Jordan, each time for about a year. The he travelled to Saudi Arabia again where he stayed for another year and a half. He has worked abroad six years in total and returned home only recently. At the time of field research he was not actively planning a future term of migration, but he envisaged leaving again after a certain period at home. From his earnings Salim first bought 1 *qirat* of construction land and built on it a house from mud brick (18). When he came back from his next journey he bought 8 *qirat* of agricultural soil, and on his following return to the village he bought another 12 *qirat*.

Hagg Salim described his life before migration as 'one long, incessant crisis', from which there had been no other escape than migration. His sole source of revenue had been the insecure, occasional daily wage labour. Had he not migrated he never would have been able to afford the money to build appropriate living space and to escape from the cramped living conditions in his paternal house, he said. His first aim had been to rigorously economize – and he was lucky enough to get a contract for work in Saudi Arabia where he could earn relatively good money. This attitude of economizing is stamped on many aspects of the couple's behaviour. He did not build an expensive, fashionable red brick house, but he used the cheapest material. He only spent a small part of his remittances on goods (the sole modern consumer durable the household owns is a black and white TV set), saving the bulk of the money to make the dream of any landless or dependent and share-cropping peasant come true: to buy agricultural soil and to escape from submission to the big landowners. He never tires of enumerating the advantages of being owner of agricultural soil:

> With half a *faddan*, but surely with 1 *faddan* of *milk* [landed property] it is possible to feed a whole family. If you plant half of it with maize, this will suffice you for the whole year, you then get flour to bake bread and you have corn to feed your poultry. You also have fodder for your cattle and the maize stalks will be dried to serve as fuel. With rice and wheat it is the same. And if the household has a *gamusa*, it can sell the milk and it will provide an earning of £E20 to 30 a week. And with this money you can buy vegetables and cereals. We also have eggs and poultry. Everything is used – nothing is thrown away. We use the inner part of the maize cobs as spools

> to wind up thread, as bottle stoppers or as coal to burn hubble bubble tobacco. We have onions and garlic. We get the rest we need, sugar or tea, from monthly subsidized food rations.

In the light of the conventional point of view expressed in the relevant literature, Hagg Salim's assessment that between 0.5 and 1 *faddan* of *milk* is sufficient to secure reproduction sounds rather modest and optimistic, given the fact that often development experts estimate that about 1 *faddan* is needed to feed a family of 3.5 persons[2] in Egypt, and that Hagg Salim's household comprises seven people. Whether or not a *faddan* is sufficient, the possession of a plot of land, however tiny it may be, is clearly at the centre of Hagg Salim's household strategies. Agricultural land, even if it is only rented on a permanent basis and even if it is only a very small plot, is a security the household can count on.

In the peasant household it is not only land and the work on it which are important. The raising of cattle and the possession of agricultural machinery are likewise crucial as providers of security. It is therefore not surprising that one of the important migration motivations for households is investment in agricultural assets, such as land, cattle and agricultural machinery. Of the questioned remigrants in Kafr al-ʿIshra, 29 per cent mentioned investment into agriculture as one motivation for their work abroad.

Table 6.1 Investment into agriculture as a motivation for migration

Motivation	*Number (%)*
Not mentioned	71
Buying land	21
Buying animals	17
Buying machines	3

Source: Remigrant sample 1987/88.
Note: Multiple responses possible.

If we take a close look at Table 6.1, purchase of land as a goal emerges as prominent, 21 per cent of the migrants stating that to buy land had been among their motivations to migrate. Migrants want to buy land, for the reasons given by a 37-year-old illiterate agricultural wage labourer:

> To buy land is much securer than to invest into a *mashru'* [a project]. Agricultural soil is like gold, it does not lose its value. The land will ensure that we can feed our family. We are also very experienced in the cultivation of the fields because we are peasants. A project on the other hand is always a risk.

The same is true, if to a lesser degree, with investment into cattle, especially a *gamusa*, a water buffalo. For the household, the *gamusa* is of great importance as it provides the household with milk, and hence also dairy products. Seventeen per cent of the remigrants in Kafr al-ʿIshra indicated the wish to purchase cattle as one motivation for migration. Finally, the possession of machinery is of significant relevance to the peasant household. Machinery is important because it facilitates the peasants' hard labour enormously.[3] Mechanization of agriculture is generally restricted to the large landowners. There are only five tractors and threshing machines in the village, all belonging to the large landowners on whom all the peasant households depend. With this background in mind it is no wonder that buying agricultural machinery is a dream of migrant households as it would reduce the dependency on large landowners and at the same time be an additional source of cash income.

Table 6.2 Spending of remittances in agriculture

Type of investment	*Number (%)*
No agr. investment	58
For animals	36
For machines	8
For land	6

Source: Remigrant sample 1987/88.
Note: Multiple responses possible.

Such, then, are the main migration motivations relating to agriculture. What, however, are the actual achievements in this respect? Are the migrants able to realize their aspirations and become landowners and cattle holders? We can see from Table 6.2 that remigrants in Kafr al-ʿIshra are able, by and large, to make their agriculture-related dreams come true. While 29 per cent of the migrants had indicated agriculture-related motivations for migration (see Table 6.1), 42 per cent of the sample realized some investment into this sphere (Table 6.2) – which is considerably higher than the number of those who had initially planned to do so. However, a closer look at Table 6.2 reveals that investment often took place in areas other than the originally desired one. Thus while 21 per cent of the remigrants indicated the purchase of land as a migration motivation, only 6 per cent of the sample were actually able to do so. Those who were successful and did in fact buy land, and for whom the households of Hagg Salim (18) and Maher and Fatima (8) are examples, usually buy relatively small plots which do not exceed a few *qirat*.[4] The reason for both the failure to buy land and the small size of the plot when acquired is easily worked out: agricultural land is extremely expensive and prices have gone up tremendously in the last years. It is perhaps interesting to note that the general opinion in the village is that the most important reason behind this inflationary tendency of land prices is migrant-demand for land.

If the purchase of land cannot be realized, then the 'second choice', and a more realistic alternative, is to buy cattle. Thus 36 per cent of the remigrants have bought animals, usually a water buffalo – whereas only half as many had previously planned to do so. A grown-up cow can be bought for £E2,000 – which is far easier to pay than the immense prices for land.[5] Possession of cattle is of great importance. A buffalo and its milk, as well as the prospects of a calf which in turn can be raised and sold, provides the household with more options to secure subsistence than agricultural machinery would do. This is even more the case as profitable usage of machinery is very often hampered by the great logistic problems owners meet when the machines are out of order. It is a difficult undertaking to get hold of required spare parts and, in addition to that, capable mechanics to repair them. Nevertheless, to buy machinery is also an option and this is reflected by the fact that remigrant households which actually buy agricultural machinery also outnumber those that had planned to do so before migration. Villagers indicated that the price for a water pump ranges between £E2,000 and £E3,000 – an approximately equal sum to the cost of a buffalo. No remigrant household has acquired a tractor or any other agricultural machinery so far. Their relatively low price explains why water pumps remain practically the only agricultural machinery these households can afford.

The strategy to invest into agriculture speaks of the hope for a future stabilization and enlargement of the base of production – which in turn indirectly aims at greater freedom from market forces.

This strategy is linked to avoiding further monetization and dependence on income from wage labour. The model for these migrants is definitely the 'autonomous' big landowning household integrated into the village society and owning enough assets to be able to sustain the family from its produce.

Investment into land most certainly has a status-related dimension. As the community is still to a large extent integrated into agriculture the amount of land a household owns appears as the most important sign of status and prestige. For those poor peasant households which have been dependent and landless since the very founding of the village in the last century, buying land is an issue of extreme importance. Being landless, dependent on agricultural wage labour, and belonging to the community of *harat al-ʿabid*, assigns one a position very much at the bottom of the community's social hierarchy. Acquiring land is of tremendous importance, then, for it secures reproduction independent from the big landowners and the market and at the same time lets one move up the social hierarchy from landless wage labourers – from being a resident of *harat al-ʿabid* – to landowners.

INVESTING REMITTANCES INTO THE MANUFACTURING AND SERVICE SECTOR

The discussion of remittance investment into an 'enterprise' in the service or manufacturing sector is quickly carried out as the number of those who spend money on such a small project is rather limited. Only 3 per cent of our remigrant sample from Kafr al-ʿIshra have invested money in such spheres, and as has already been made clear in Chapter 5 none of the migrant households in the two streets in *harat al-ʿabid* was able to do so.

The three remigrants of the sample who actually did invest into small enterprises outside agriculture have done so in spheres which are bound to the accelerated speed of monetization in the countryside. These projects are generally an expression of transformations which have recently occurred in the wake of *infitah* policy and its stress on the service sector. People are getting more mobile and thus new and more means of transportation are needed. Two of them have bought a pick-up which is used as a sort of communal taxi on the asphalt road connecting the district towns of Mit Ghamr and Sinbillawein. The third one has opened a carpenter's workshop on the asphalt road. As new urban-style furniture has found its way into the village and is today considered a 'must' for any couple about to marry, this remigrant has seen a chance to make a living by turning from the production of the traditional bridal wooden trunk to Louis-XVI-style sitting rooms. But he also produces doors and windows and thus is able to profit from the recent construction boom in the village. Though not part of the sample, I know other remigrants who have invested in such small, often service-related enterprises. Also situated in Kafr al-ʿIshra's 'industrial zone' is a welding workshop. Its customers are mostly owners of cars and agricultural machinery. One other man has opened a small *butik* where his sister sells Coca-Cola, Fanta, cigarettes and sweets. Another man has bought a truck with which he transports construction materials and agricultural products. With the exception of the *butik*, most enterprises demand considerable amounts of capital and experience, basic requirements which are not very often met with in Kafr al-ʿIshra. Migrants from this village generally do not earn much money abroad, and the sort of jobs they perform don't allow them to gain new experiences that they could capitalize on when back in the village. Only 14 per cent of the remigrants stated that they had gained new qualifications and experiences during their work abroad.[6]

Service-oriented projects directly integrated into the market economy are still rare and find their way into the countryside slowly. In Kafr al-ʿIshra spending of remittances still takes place predominantly in the sphere of agriculture. Migrant households in Kafr al-ʿIshra prefer investment into what appears to be secure and known – into agricultural production, and into the consolidation of the household as a unit of production and reproduction – rather than risk losing scarce funds in off-farm projects whose fate is considered as insecure.

Remittance spending in off-farm activities (e.g., in the possession of a car or a lorry) has a symbolic aspect, too. While the investment into land may be interpreted as a symbol for an orientation towards 'village traditions', having a car is clearly a symbol of a modern lifestyle and for participation and integration into the outside world beyond the narrow village boundaries.

REMITTANCES AND THEIR EFFECT ON LIFE

The economic and subsistence-related character of remittance spending is less obvious when it comes to housing, education, or medical treatment, etc. The question arises as to whether spending money on such assets or services is still 'productive' or rather 'consumptive'. In this sense, many writings stress the predominantly 'consumptive' remittance spending of remigrants (Abu Lughod 1989; Adams 1986, 1988) and deplore the 'squandering' of scarce capital which could have been better invested into some small income-generating project. Yet I hesitate to put the blame on the migrant household and accuse it of being victim of the enticements of advertising and TV with its propagation of a consumptive life-style. Several arguments may be forwarded in this respect. Firstly, in most cases money earned abroad is not sufficient to establish a viable 'project'. Lack of capital, skills and consultation also have the effect of keeping such projects outside the reach of most migrant households in Kafr al-ʿIshra. Another aspect is that the living conditions of many poor peasant households are so poor and precarious that even the purchase of such items as blankets, fabrics and clothing appears to be an investment into the production and reproduction of human life in a wider sense. It is with these considerations in mind that I would suggest subsuming investment under the following items and services as being necessary for the continuance of life and hence directly related to the economy of the household.

Let me first of all consider the spending of remittances for medical treatment. Mentioned by 20 per cent of the sample (see Table 6.3), it reflects a great concern to be able to visit a private doctor or to buy medicine if necessary (both are rather expensive for the Egyptian lower and middle classes). The alternative is to take recourse in methods of traditional healing, which are not practised very much nowadays, or to direct oneself to the public health care institutions. Though principally within the reach of every individual, these centres are often awfully overcrowded and 'have nothing more to offer than aspirin', as one woman said. The consequence very often is to do nothing and 'to leave it in God's hands'. I have been shocked by the great number of women especially who suffered from severe illness and were not able to afford the money to consult a doctor and buy the prescribed medicine, let alone purchase enough for regular, long-term treatment. I also have already drawn attention in the last chapter to the extremely high incidence of infant mortality. It is therefore indisputable that very much has still to be done to secure at the least very basic and rudimentary medical care. It is with this situation in mind that the spending of remittances on medical treatment becomes really significant: it reflects the inability or neglect of the government to put sufficient, affordable and accessible basic medical services at the disposal of those who are in greatest need for them – the rural poor. Spending remittances on medical treatment is not a luxury but a very basic necessity to sustain health, often even to prevent death – hence it is directly related to the production and reproduction of life.

Table 6.3 Use of remittances for general purposes

Use of remittances	*Percentage*
Electrical consumer articles	91
Building house/flat/floor	53
Furniture	35
Education of children/siblings	20
Migrant's marriage	15
Medical treatment	20
Pilgrimage	6
Construction land	23
Gold	12
Animals	36
Marriage of children/siblings	4
Agricultural land	6
Agricultural machinery	8

Source: Remigrant sample 1987/88.
Note: Multiple responses possible.

Let me now consider various forms of investment which are all more or less directly related to marriage. First of all, a total of 53 per cent of the remigrants spent money on the creation or amelioration of living space. People build a new house, an extra floor or additional rooms. Twenty-three per cent of the remigrants had bought construction land with the aim of building a new house, and 14 per cent had used remittances for the general improvement of the house, for example for the whitening or colouring of the walls. The importance of remittance spending for the procurement of living space is easily understandable given its general scarcity in Egypt. This 'crisis' is not confined to urban areas, but is increasingly finding its way into the countryside. Reasons for the extreme population pressure are to be seen in the rapid growth of population[7] and the scarcity of construction land in the main residential areas – new houses practically can only be built on the scarce agricultural soil.[8] Today there are also more young people who realize their dreams of separating from the extended household and founding their own nuclear family. The 'housing-crisis', even in Egypt's rural areas, clearly has a reproduction-related dimension, as without at least the availability of a separate room for a young couple in the paternal household, marriage, the precondition for procreation, cannot take place. But even the providing of a single room in the house of a small peasant family often poses problems.[9] Peasant houses do not usually comprise more than three, at maximum four, rooms, which is hardly sufficient for the parents and unmarried children themselves, let alone married sons and their wives and children. Thus constructing additional rooms or an additional floor – and sometimes even a separate house – is an absolute necessity for the continuance of life in the village.

A related expenditure is that for furniture: 35 per cent of the migrants reported that they had spent money on such items. These expenditures are usually related to marriage, as furniture is still a lifetime investment that is to be purchased before marriage. On the other hand the buying of furniture is a clear example of the difficulty in deciding what is to be understood as 'needed' for subsistence and what is to be defined as luxury. Here, standards and values definitely have changed in recent times. According to the villagers only some very basic equipment was needed for marriage 'in the past', while today velvet upholstery, a dining-room and a kitchen are required. And those who have remained below this 'standard' are talked about in the community as having prepared a marriage *ayy kalam* – approximately a cheap, too simple marriage only for appearance's sake. What is needed for an adequate marriage unquestionably has a cultural dimension, here as anywhere else. One can discuss the indispensability of an ever-growing set of household items, equipment and utensils as an instance of market integration and increasing monetization, of the introduction of urban life-styles into the village – but that does not change the fact that a certain clearly defined set of items is needed as a precondition for marriage and the resultant continuance of village life which it brings. Households which have not been able to acquire sufficient 'items' will face enormous difficulties to marry their children.

Education of children or siblings, mentioned by 20 per cent, is an important issue for the remigrant household, too. It is a strategy which is definitely important as a way to secure the future of children: 'School education is a good thing and a graduation

Plate 6.1 Fashionable Louis Cairo upholstery. The woman on the left wears a black dress because she is in mourning. Her husband had died from the late effects of bilharziasis. The woman on the right – the owner of this house built in red brick – wears an urban-style nightgown. Both women are neighbours. The table is decorated with a doll, actually containing a dish for confectionary. This is a reminder of the time when she worked in Cairo. The material under the table is used to cover the upholstery and had been removed especially for the photograph.

certificate guarantees the future,' as one remigrant (28, carpenter, preparatory school not finished) put it. Another migrant said:

> If I were educated, then in times where we cannot do wage labour in the fields, I could find a job at any other place. But now there is no work. If I had been educated, then I could have found a hundred jobs instead of spending my savings from my work abroad. That is why it is so important to send our children to school and to buy what they need for it. And the migrants with much money should build a school, so that our children need not go to the neighbouring village.
>
> (Remigrant, 32, can read and write, wage labourer)

Though basic education is nominally free of charge many of the parents in the two streets of *harat al-ʿabid* complained that they could hardly afford to pay the fees for books, clothing, pencils and the like for all of their schoolchildren. What poses even severer problems to these families are the costs of private lessons. Attending such private lessons comes near to being obligatory for those pupils who care for education and want to continue secondary school.[10]

As the two remigrants indicate, it is no doubt that education is an investment into the future of the children, and hence into the quality of labour power. Their opinions clearly mirror the generally held views of large segments of the rural population of Egypt, and perhaps those prevailing in most regions of the Third World as well. But it can also be observed that migration may have a negative impact on education. This is indicated in the following observations of a primary school teacher:

> Migration of the fathers clearly has an impact on children. Many parents say they migrate because they want to secure their children's future by giving them a good education, but the pupils often see things differently. Smaller children lack the authority of the father to make them concentrate on their homework and the older the pupils get, the more they adopt an indifferent attitude towards learning. I think they do not see any use in reaching higher education because of Egypt's employment crisis. They say: 'What is the use in all this drudgery – I want to work in the Gulf states and will make much more money there than I could ever earn here, even if I had a university degree.'

In sum, the spending of remittances on the creation of living space, on furniture and on education, is more or less directly related to marriage which in the Egyptian sociocultural context is the absolute precondition for procreation. I therefore regard financing a marriage as an investment into the reproduction of life. Without doubt conditions and preconditions of marriage have considerably changed over the last years: bride-prices – be it for a *mitʿallima* ('educated') or for an uneducated peasant girl – and the value of the required dowry have risen tremendously. In this sense marriage is one of the outstanding examples of the incessantly rising influence of money on the countryside. Everybody is lamenting over these developments as they constitute serious impediments for marriage. As ever-longer periods of wage labour are required to afford the necessary sums, the age of marriage is also on the rise. This has led to the prevailing opinion that the one who wants to establish a family has to work abroad for a year or two prior to marriage:

> then my son can go abroad and perhaps improve his situation. As long as he stays here, he cannot do anything: he cannot marry or build a house. Here, he can do nothing at all.
> (An agricultural wage labourer, 37 years old, illiterate)

The need to migrate in order to marry has led to the effect that migration is now a distinct period in the life cycle of a young man, a reality which finds expression in the following quotation:

> Of course a young man who is still at the beginning of his life, and who has no chance to marry, should migrate. Instead of sitting here and waiting for some years for a government appointment he can migrate. Here he would only do a job which will not earn him enough. Therefore he should go to meet the marriage expenses and to stabilize himself. He migrates until he gets his appointment and in the meantime he has built a house and his situation has totally changed.
> (A cooperative member, 30 years)

Returning to the families of the two streets in *harat al-ʿabid* (see Map 5.1), we have already come across a young couple which migrated before marriage (9). While she had been to Cairo, her husband had worked in Iraq. In what follows I want to present the case of Ibrahim (1) as it illuminates many of the points made above.

The case of Ibrahim

Ibrahim is 25 years old and is married with a little daughter. In 1982/83 he graduated from an agricultural institute and obtained a diploma, but he has not yet been appointed to a government job. Together with his wife he is living in his father's house. There is still an unmarried sister staying with them. The household does not have land, nor a *gamusa*. The only source of cash income is agricultural wage labour, performed by all the family members except for his wife. His mother also works in the household of a rich landowning family and is rewarded in cash and kind. The first time Ibrahim went to Iraq to seek employment was shortly after his engagement in the summer of 1985. He stayed abroad for one year to earn the money necessary for marriage. When he came back he had saved about £E2,000:

> I bought the *shabka* [the bridal gold] for £E900. I brought with me from Iraq a dress for my bride for about £E100. I have paid £E2,000 for the *mahr* [dowry] and with this money we bought the sleeping room and the living room. The parents of my bride bought the dining room and the kitchen appliances. I have spent about £E4,000 to marry. I could not have done that without the help of my father. He sold the *gamusa* for me. I have bought the sleeping room for £E1,200, the living room for £E600, a black-and-white TV for £E150 and a radio-recorder for another £E95. I built two rooms on the roof of our house from mud brick as a home for me and my wife. But we eat and drink with my family, because until today I have no job. My wife is educated, she has a diploma like me. If I had not gone abroad, I would never have been able to marry and I would have had to take a poor *fallaha*, who would have been content with an iron bed and a wardrobe.[11]

After his marriage Ibrahim stayed with his wife for ten months, when she gave birth to a daughter. Then he left again for Iraq. This time he stayed only eight months. He came back earlier than he had planned because he did not find adequate work, as he said. In summer 1988 he was preparing for a third journey. Ibrahim has put it quite clearly: without migration, he could not have married. The reproduction of life needs as a first prerequisite marriage, and marriage in turn needs the prior securing of housing. With the help of his parents who had sold the *gamusa*, their sole productive asset, and with a stay abroad, Ibrahim managed to marry. His immediate future is marked by a new term of migration in order to secure the sustenance of his young family and that of the extended household of his father. He does not see any other choice open to him than to migrate again. The case of Ibrahim exemplifies the results of the statistical data, according to which 15 per cent of the remigrants in Kafr al-ʿIshra had spent remittances to cover marriage expenses, with 4 per cent indicating that they had done so for the marriage of their children or siblings (see Table 6.3). Marriage-related remittance expenditure is also hidden in other statistical entries, as I have previously shown. Investment into marriage is most probably also the rationale of many of those 12 per cent who bought gold.[12]

At this point let us summarize the main findings about remittance spending which has the ultimate aim of the reproduction of life. The overwhelming majority of migrants from Kafr al-ʿIshra return from abroad with only modest remittances in their pockets. Consequently, prospects of successful investment into income-generating projects and activities remain rather limited. Migration motivations are often linked to the wish to enlarge the household's agricultural base of production. Migrants from small peasant households in Kafr al-ʿIshra want to buy land. But only a minority realize this hope. Their failure is partly due to limited resources and partly to the dramatically rising prices of agricultural land. Alternatively, money is invested in a water buffalo, a very important productive asset in the peasant household's economy.

Migration motivations and the spending of remittances in Kafr al-ʿIshra underline the strong agricultural orientation of its migrants – who are thus in concordance with the village population in general. Rather than invest in an off-farm activity, they spend money on the purchase of cattle and land. While the hope to fundamentally improve the household's economic situation remains unfulfilled in the overwhelming majority of cases, remittances are often sufficient to achieve some very basic amelioration of living conditions and a certain reinvigoration of the household's base of production. It is in this sense that the spending of remittances in such domains as the creation of living space, marriage, better nutrition, medical care, and education can be understood. This kind of spending is no luxury, but necessary for the production and maintenance of the household's labour force.

In conclusion, it is the basic amelioration of the conditions of agricultural production which also secures and perpetuates the small peasant household's functions within the global system: the peasant migrant households themselves, by turning migration into a strategy and by the purposive investment of remittances, secure their maintenance as a pool of cheap migratory labour.

SOCIOCULTURAL ASPECTS OF THE SPENDING OF REMITTANCES

I now come to the argument that migration is not undertaken just for strictly economic ends but that other strategies may implicitly or explicitly be pursued as a result of it. It is these non-economic purposes I want to deal with now. Two aspects of non-economic forms of investment may be discerned. One form is status related, examples of which have already been given. The other involves what one may term the overcoming of repressive living conditions.

Issues of status and the desire to be part of a modern life-style – that is, being able to display its symbols – are always inseparably linked to any kind of remittance spending. Often it is because of the very fact that someone has been abroad to one of the rich Arab countries that a person has got acquainted with the 'wide world', with the manners and habits of a foreign people, with the knowledge of how to behave in an airport and a plane, which bestows on this person a higher status. Another very obvious example of such a 'secondary effect' of migration is the performance of the *hajj*, the pilgrimage to the holy places of Islam in Mecca. Many Muslim migrants who work in Saudi Arabia comply with this religious duty during their stay abroad. When they come back home they have added to their name the honourable title of *hajj*,[13] a new status which finds expression in the wall-paintings of religious content which relatives paint in the expectancy of the migrant's return.

These non-economic forms and impacts of remittance usage are also of particular interest when it comes to the purchase of the so-called modern mass-produced consumer durables. Outsiders to the village, whether development planners or social researchers, often denounce them as purely consumptive, as a waste. However, the use of such appliances has to be more closely observed because they may often be used for purposes quite different from those which the manufacturers have anticipated. Still other non-economic migration motivations are related to the individual's interest to overcome restrictive living conditions. In many cases such aims are well hidden behind an alleged economic purpose. As such personal goals may be much more sensitive to the involved persons than the pursuing of economic aims via migration – which is even expected in the village society – they may find no open expression. While this is an issue also inherent in marriage, I will illustrate my argument by taking the case of household separation.

Uncovering such motives is also important for theoretical reasons. Understanding the hidden incentives for migration is important inasmuch as they undoubtedly demonstrate that migrants are not necessarily the victims of 'market forces' pure and simple, and are driven not only by exclusively economic rationales but indeed are actively working for their own aims and for a positive change in their practical living conditions. Agents, whether migrants or not, use structures for their own individual ends. This is, in brief, my guiding line for the exposition which follows. But first, a look at the spending of remittances on modern consumer durables.

Remittance spending on consumer durables

Almost every migrant household (89 per cent) in Kafr al-ʿIshra has bought long-life consumer durables. Here, the most important item is definitely the TV set, mentioned by 82 per cent (see Table 6.4). The necessity to possess a TV set, and its usefulness, are unquestioned in the village. The most positive effect of TV to many people here has been that it has contributed to the opening of the outside world to the village and has made it possible for any person interested in participation to get in touch with it. This is so for a variety of reasons. Most villagers' movements, and especially those of women, are still restricted to the village boundaries or at most to its nearer surroundings; most of them are illiterate and moreover, printed matter, be it books or even newspapers, does not have a market in the village, people being too poor to buy them. TV has the important function of bringing modernity – that is, *tatawwur* (development) to the village. Here is an example.

Hagg Salim and Aisha's (18) 1-year-old daughter had been ill for about two weeks. She suffered from diarrhoea, high temperature and listlessness. They had not taken her to a doctor but had bought some medicine: 'we know which medicine is prescribed in such cases'. Aisha and I discussed what would be the best diet for the ill girl and she told me what she feeds her. She had bought a packet of baby food. She mentioned that she knew from TV what to do – everything must be clean, that she has to use boiled water and clean glasses. She went on to elaborate on the great importance of TV in this respect. Without TV, getting this kind of information was very difficult simply because nobody was aware of these facts; ignorance would lead to children dying. But now, with TV, things were getting much better then before, she concluded.

It is also through TV that villagers have got an idea about life in western countries, and it was this knowledge they introduced when discussing with me differences between our ways of life: in my country everything was clean, organized, beautiful and easy, there were the *hagat al-helwe* – the 'nice' or 'beautiful' things they did not possess. Their own life was retarded in so many respects, it seemed. For many men the most important transmissions are football matches, with those involving Egyptian teams being only marginally more popular than other matches. Often, men were quite disappointed about my complete ignorance in this respect, as they wanted to discuss with me the faring of German football teams. International (mostly Indian) and especially Egyptian soap operas, are also continuously followed.

The radio-recorder is second in importance as a long-life consumer durable: 52 per cent of the remigrant sample had bought one. It appeared to me that the most regularly listened to radio programme was probably the daily broadcasted songs of the famous singer Umm Kulthum, now dead. This preference seems to be widespread, for, walking through the streets of Cairo after six in the evening, Umm Kulthum's voice resounds from many small commercial enterprises and workshops, uniting all Egyptians into one large national community. As a medium for the transmission of information, such as the news, women's programmes, religious instruction and the like, radio falls behind TV. Yet radio keeps the village youth up to date with Egyptian pop songs. More important in this respect is the integrated cassette recorder. Cassettes are usually too expensive to allow the village youth to buy all 'top of the pop' recordings of Egyptian and foreign singers which appear on the market. Most of them only possess a quite limited number. Generally speaking, Egyptian songs are very much favoured over those of any other Arab or especially *agnabi* ('foreign', i.e. western) pop stars. Knowing and possessing such Egyptian and western music has an important symbolic aspect, too, as hereby familiarity with and integration into the 'modern world' are demonstrated. Here is an example. I went to see Sahar (7) to congratulate her on the occasion of an important Muslim holiday. I found a youth of about 18 years at her home, a relative living near Alexandria. Sahar's radio-recorder was playing the Egyptian pop-song cassettes he had brought along with him at full volume. Sahar was happy and joined in singing the songs she liked most. While initially having been dressed in a *gallabeyya*, soon after my arrival the youth disappeared and eventually returned in jeans and shirt. He wrote down a list of the most famous and up-to-date Egyptian pop songs so that I could buy them when going to town. He then apologized with the words 'If I had known in advance that you were here I would have brought along with me my foreign cassettes, too.'

Batta, the Christian woman we met earlier, had found that the cassette recorder was useful to her in still another way. She used it to wake up her husband in the mornings, a procedure usually stretching over an hour. To achieve her aim she always used the recording of a Coptic sermon held in one of Egypt's famous churches. Starting with the church bells ringing she put the cassette recorder on full volume. As all these examples pointing to the multi-functionality of both TV and radio-recorder, it is hard to accept them exclusively as a form of consumption.

Table 6.4 Remittance spending on consumer durables

Item	*Percentage*
Radio-recorder	52
TV	82
Washing machine	59
Private car	1
Refrigerator	6
Fan	15
Mixer	4
Iron	9
Sewing machine	4
Butagaz	13

Source: Remigrant sample 1987/88.
Note: Multiple choice possible.

Other frequently purchased items from remittances are of more evident practical utility. Here the washing machine (59 per cent) ranks highest, followed by a fan (15 per cent) and the *butagaz*, a gas stove used for cooking (13 per cent). Migrants in Kafr al-ʿIshra have bought an average of 2.45 items in the category of 'consumer durables' (i.e., electrical durable appliances). Anybody who knows the working day of an Egyptian peasant woman can hardly postulate that the investment into a washing machine or a stove are signs of 'conspicuous consumption'. Without doubt these items are status symbols and give proof of the household's integration into the modern world. But to argue that such expenditures are useless and should have been avoided in favour of the establishment of income-generating projects appears to me as rather cynical, misleading and even misogynous.[14]

Remittance spending, marriage and status

Let us once more turn to migration and its interrelationships with cultural dimensions of marriage. As the *mitʿallima*, the educated wife, is a sign of a household's high status, girls with a formal education are desired brides. But these are far more expensive than a 'poor (and uneducated) *fallah* girl', and therefore migration is all the more necessary if one wants to marry a 'suitable' bride. The payment of the dowry and *shabka* is an indispensable condition for a marriage in the village, yet the quality and quantity of these bridal rights have changed considerably over the last years. Villagers often see the reason for this recent development in migration. Today migration and marriage seem to be linked in a kind of vicious circle: without migration, marriage is not possible, but at the same time migration and the ensuing monetization have caused a dramatic rise in bride prices (see also, Abaza 1987: 90):

> Everybody who wants to marry today spends very much money. Since migration *mahr* and *shabka* have become very expensive, because migrants come back from abroad with a lot of money. For the illiterate and the educated, *mahr* has become very expensive today, because everybody looks at his neighbour .... In former times *mahr* was not more than £E300 or 400. Today one must pay £E2,000 to 3,000.
>
> (A remigrant, 28 years old, married)

'Inflationary' tendencies are observable not only with regard to the bride price, as today an ever-increasing number of assets are needed as a precondition for marriage. Let us listen again to Ibrahim (household 1):

> In former times they bought an iron bed and a wardrobe. The mattress cushions and covers were made from cotton. The bride got copper pans and pots, a gas light and a *babur* [a gasoline cooker] and dresses. That was all one needed. It didn't cost more than a hundred pounds. In former times people did not have a dining room, they did not even know what it was .... Today we need the furniture for a whole sleeping room, for a living room and a kitchen. Instead of a *babur*, the bride demands a *butagaz* [a gas stove] and she wants a washing machine and a TV set and *sini* ['Chinese' porcelain]. All men, whether they have migrated or not, have to buy modern furniture and clothes for the bride. The non-migrants here face real problems.

Ibrahim, the holder of a diploma, married an educated girl and so consolidated the high status his education had already given him. Being educated, then, clearly has a status dimension. This can already be deduced from the widespread stigmatization and construction of the 'peasants' as 'traditional' – as ignorant, illiterate, uneducated, stupid and emotional.[15] In this sense to get a *mitʿallima*, an educated woman, as a daughter-in-law clearly indicates status.[16] The most important mark of an educated bride is the amount of gold[17] she gets, thus making remittance spending even more important.

It is very clear that today marriage has become one of the obvious vehicles for the introduction of urban life-styles into rural society. The establishment of a young family, which is the very basis of the reproduction of life, is imbued with the symbols of modernity and urbanity. This is already apparent during the marriage ceremonies and festivities. It is very interesting that these festivities are closely linked to the notion of *nadhafa* ('cleanliness'). *Nadhafa* is the opposite of *ʿaraf*, the things that 'disgust' in the village: mud, dust, flies, lice, mosquitoes. Cleanliness is where one eats with knives and forks, where one goes to the hairdresser's every week and has one's nails manicured, the young girls joke. Brides are always dressed in the white, foreign bridal dress, meant to symbolize their purity, their cleanliness. Yet here the white colour also signifies distance from the 'filth' of the village and the 'earthiness' of its inhabitants, and distance from being of 'slave-origin'. Slaves were black, as we have seen, and though there are not many black people in the village today, many of them have quite a dark skin-colour.[18] Brides, of course, wear the heavy make-up of many middle-class urban women, yet what struck me here was that their complexions had almost been turned to white, sometimes nearly chalk-white. Brightness, the illumination of the traditional village at these joyous, important occasions, is given further expression on the 'henna'-night, the celebrations on the night before the bridegroom takes his wife home. Parents who can afford it, install a great number of neon tubes in their alley, the place where villagers come to congratulate the bride and her family. All these marriage preparations, of course, force expenses up and have to be covered by remittance spending.

Marriage expenses, though, are driven up at a much earlier stage when a girl's parents struggle to afford an education for her which will allow her a certain upward mobility through marriage – almost the only strategy poor people can employ in this respect. It is here that the notion of *nadhafa* comes in: mothers quite often say that through educating their daughters they want to secure a better future for them by making it possible for them to have a marital life in a 'clean' urban surrounding, far away from the village's *ʿaraf*.

Another instance of the impact of migration is the gradual breaking up of the household's traditional power structure. Let us once more listen to Ibrahim:

> In former times the bridegroom's father or grandfather chose the girl he was to marry. When the grandfather was still alive he had the last word because he was the owner of land; if he did not possess any land then he was the senior person in the family and everybody had to listen to him.
>
> At that time the one who was about to marry only had to say 'yes', after than he had no responsibility whatsoever and his father and his grandfather were responsible for everything, also for the bride. The father paid for the *mahr* and the *shabka*, because the money was in his hand, and he also had his son in his hand because he was living with him before and after marriage. The father decided what the son had to do. After the marriage, he had the son's wife and children to work on the land and took away their money. Today things have changed. The educated migrant chooses his bride on his own. He no longer listens to his father. He chooses a girl, even if she is not from the family or even if she is from another village. He is free because he pays the marriage expenses on his own. When the migrant wants a different girl to the one his parents have chosen, then he says to them: 'If you do not agree, then I marry her all the same' and then the parents have to give in. Today anybody can marry the girl he loves.

The statement of Ibrahim very clearly shows how sons who have got hold of substantial sums of money through migration are today able to ignore their fathers' hitherto unquestioned authority (see also Evers and Schiel 1981: 291, Kandiyoti 1977). Fathers are no longer the sole holders of economic assets and the only persons with the privilege of decision-making. The father's right to choose a daughter-in-law according to his own interests – for example, in an additional input of labour power and source of cash – has gradually been questioned because migration has given sons an economic position independent from their fathers. It is evidently a side effect of monetization which gives the young migrants the chance to 'revolt'. Possessing his own source of cash, the migrant son claims for himself the right to choose. Probably the young generation's wish to realize its own aspirations against the will of parents has always existed, and it is only through migration that they have gained the economic power to have their own, independent way. It is interesting to note that the wish to choose one's bride independently is also accompanied by new ideas about what an ideal marriage is. Today the notion that a good 'match' is the one which contributes most to the household's base of reproduction has increasingly to compete with the importance of love as a reason for marriage.

The father, in turn, cannot publicly oppose his son's wish to work abroad, because migration is generally viewed as almost the only way to realize a marriage. If he did, the father could be accused of hindering his son from performing one of the most important rites of passage in his life, an institution which is deeply rooted in the village society's value system and norms. Thus the son's migration and his marriage project must appear to be the decision of the whole household as a unit: a son has proceeded to a certain stage in his life cycle where marriage and having children is what 'naturally' comes next. However, if we have a closer look at intentions and facts we realize that the son has found in migration a way to exercise his own interests and to make his own dreams come true, even against his father's interests and point of view: 'Today anybody can marry the girl he loves.'

The preceding discussion provides evidence for my argument that households are not a priori homogeneous entities, but that their members may follow quite different and conflicting interests. To some extent the marriage of a son is the undertaking of the whole household unit, which has to allocate resources and space for this purpose – and gets in return additional labour force. In this sense it is the concern of the whole household. Yet, when money gets into the hands of the son this gives him considerable freedom to select according to his own choice and to ignore the will of his father and mother.

Marriage seems to be only a first step in such an 'emancipatory' process of the sons, which is eventually followed by other, equally significant, ones – and here again the spending of remittances comes in. After marriage ceremonies young men stay for a while in the paternal household together with their wives. Eventually they will depart again, leaving their spouses – pregnant or with a young child – under the shelter of the father. One important objective of this renewed travel, as perhaps of others to follow, is the accumulation of an amount of money sufficient for the purchase of a *qirat* of building land and the construction of a house for himself and his wife and children. Household separation, then, is another step by which migrant sons escape from their fathers' authority. This tendency of 'emancipation' from the fathers through spending of remittances on housing is clearly visible in the two streets in *harat al-ʿabid*. Here the separation of households is no isolated phenomenon, as out of the eleven nuclear households, nine[19] have been established through the investment of remittances. Young families often give as a reason for their separation from the extended household of the father that they felt cramped in the paternal house. But another ulterior motive can be discovered. Migrant sons often work abroad to escape from power relations within the extended family. For the sons[20] this is not necessarily an abstract matter or the desire to adapt to new values and to emulate new norms. The constraint to bow before the father's will is very real and concrete and often to be felt in their practical everday life:

> My father always took the money from me and my wife, and we could not buy anything without his consent. We only worked for our food. When I wanted to buy a *gallabeyya* I could not do that because my father provided us with new clothes once a year. He always wanted to save money in order to buy land – that was the most important thing for him . . . . My father was responsible for everything in the life of my wife and me: for food and drink and sleep. The whole day I had to work in the fields of others. Everyday my father used to search for jobs for me and my wife and in the evenings we had to hand over the money to him . . . . I was living like a donkey, working the whole day and in the evenings I only ate and then went to sleep. When I wanted to stay in bed a little longer in the mornings, I was insulted by my father . . . . Therefore I migrated. The advantage of life in the separated household is that the man alone is responsible for his home. He can buy what he wants; he can also eat and drink when and what he wants; and decide whether he leaves to work early in the morning or not. Nobody commands him . . . . My wife can dress the way she likes. And I can invite my friends to my home . . . . I migrated because I wanted to separate from my family. Therefore I went abroad to save money and to buy a house for me and my wife. And I also wanted to buy furniture and electrical appliances.
>
> (A remigrant, 29 years old, married, illiterate, Kafr al-ʿIshra)

Fathers in turn fear that their sons' migration leads to a permanent loss of power and status. The migrant son's absence from the paternal household may in the end turn out to have been not temporary but for ever. Besides the decline of the available potential of the household's work-force, the loss of authority is also feared by the senior family head, as the aforementioned migrant observed:

> The father wants his family always assembled around him. He wants his family to be large and strong. If a son moves out, then the father is afraid that he can no longer decide anything. He wants everyone to live together, so that his land and possessions and the products of his land might not be divided.

The quotation clearly exemplifies the interest of the paterfamilias to accumulate and to keep together productive assets and, what is more, to control them. Labour power is an important resource of the household. Therefore migration and the probability of household separation are clear threats to the head of the agricultural household and its endeavour to secure the family's reproduction. However, the paterfamilias is not deprived of mechanisms to protect his interests, as his position is still backed to a large degree by the *taqlid al-qariya*, the 'village traditions'. As a man from Kafr al-ʿIshra explained when asked for the reasons why villagers stick to life in the extended family:

> If a person leaves the authority of his father – that is, his house – then there is a lot of gossip. People say: look, so-and-so has left the house of his father, his father could not discipline him. If he leaves his father's authority then he will be deprived of his heritage . . . because they say in the village that only the one who stayed on the land and in the father's house is entitled to inherit.
>
> (Peasant, 39 years old, illiterate, non-migrant)

Household separation is thus not an easily committed step by a son, as a father's authority is backed by the moral code of village traditions as he controls land. The disobedient son risks being blamed and looked upon with disapproval by neighbours and relatives. Furthermore, the father can also threaten to hold back the son's heritage – a powerful weapon in his hand.

Migration as a strategy for women left behind

If migration is a strategy of men, so it is for women. This is the case even when they themselves do not leave the village, for in an indirect way women make migration a strategy to reach their aims. And it is not only their husbands who see in it a way to escape from suppressive living conditions.

The situation of the young daughters-in-law is often far more precarious than that of the sons, as they have come as strangers to their husbands' households and they often remain so. In the hierarchy of the household the young bride ranks very lowly and she is only able to enhance her status when she has become a mother. Most young daughters-in-law complain about the fact that they are under the complete authority of their mothers-in-law, that they have to work day and night and always have to do the most menial tasks.[21] The low status and the harsh exploitation of the daughter-in-law's labour power are no secret in the village society, and their treatment is even accepted by the elder women themselves. There is in fact hardly anyone who is concerned about the young woman's destiny other than the woman herself. Women, once having climbed up in the household's hierarchy, remember their own first years of marriage seemingly without a grudge. They laugh and tell the stories about how many times they had been *ghadbana*[22] ('to be angry') and went back to their fathers' households because of quarrels with husbands or the need to seek relief from bad treatment by in-laws. Taking refuge with the father's family and reclaiming his protection is probably the only publicly accepted right of the young woman to defend herself. But being *ghadbana* is usually only a temporary relief for the women, especially if they are Christian (as there is no way for the Copts in Egypt to divorce). Eventually the father of the *ghadbana* will urge her to go back to her husband. With migration, however, women have found a new chance that they can employ to their advantage.

The case of Helwe

This is the case of Helwe, a young and beautiful woman in her early twenties. She is married and has two small children. I want to present her case in some detail because she is a typical example of a village woman in two respects: she shares the fate of having been married without having been asked for her consent, and she possesses the widely respected and esteemed qualities of *al-mar'a ash-shatira*, 'the clever woman'.[23] She is living in the household of her parents-in-law, who own 3 *faddan* of agricultural soil. The household is composed of the senior couple, two married sons with their wives and children, and two unmarried sons. Helwe's husband is a government employee on a chicken farm and earns £E50 a month. In the afternoons he works on the fields of his father. Helwe also works in the household's fields. The household is Christian. She says:

> You know, no Christian woman in our village is asked whether she wants to marry this man or that. She is confronted by her father with the fact that she has to marry now. And then you have to get along with that man somehow as long as you are alive. Divorce does not exist . . . . What can we do? This is our custom . . . . You see, I am one of these women. I was 12 years old when my father came to me and said: 'You will marry your cousin now!' And then I got engaged. They did this because at that time my aunt went blind and she needed someone to serve her. I had to leave school . . . . But I do not blame my husband, as he, too, had no other choice than to marry me . . . . Today we have accepted each other. We had no other choice. Our problem is that we have so little money, because his job is remunerated so badly, and still he has to hand over half of the salary to his father. I try everything to get hold of some money. I keep rabbits in the courtyard. I now plan to raise a larger number on the roof of the house, where there is more empty space. I go to sell them in the market. The *gamusa* is also mine. I sold my *shabka* [bridal gold] to buy it. Every morning I go to sell the milk in the neighbouring village. But my mother-in-law always demands that I give my money to her. It is difficult for me to save money. I try to economize, because we have a plan: I have talked to my husband about the *higra* ['migration'], and I persuaded him that we must give it a try. Then we can build our own family and separate from my parents-in-law. With the money I save, we can cover travel expenses and when he comes back we will build our own home. I talked about that with my parents-in-law, and they agreed to his departure after I had told them that we will not move into a separate house but that we will build a second floor on the roof and that I will go on to serve them, working in their household and in my own at the same time. You know, once we have established facts and I cease to go downstairs, what can they do then?

Helwe actively prepares the migration of her husband in order to escape from the 'over-exploitation' of her mother-in-law. She is prudent enough not to go into open opposition to her in-laws or to her husband. But she carefully worked out a long-term strategy to reach her aim. For her, living in a nuclear family, together with her husband and children, is definitely preferable to a life in the extended household of her parents-in-law.

But it is not only the workload which prompts women to strive for the establishment of a separate household. Many husbands, in the conflict between the women, side with their mother rather than with their wife.[24] In these cases women see in the establishment of a separate household a chance to turn the balance of power to their favour. There are also migrants who clearly expressed that the constant quarrelling of the women had been one reason which had led them to migrate, as they saw no other way to solve the incessant conflicts than to move out of the paternal household. The husband of Sahar (7) is a case in point. Here the mother-in-law is not his own mother, but the second wife of his father:

> After my marriage we stayed with my father and his wife. Many problems arose between my wife and the wife of my father – that is why I had to leave the house and go to work abroad. The reasons behind my travel were definitely the quarrels between me and my father and also that his wife always upset my wife. Later I got my share of the heritage, even though my father was still alive – we received three rooms in his house. We separated from them by erecting a wall and an independent entrance. My wife and I worked as daily labourers. But the money did not suffice and thus I went to Iraq. Today I do not have any problems. If there are many women living in one house, they always make problems.

I have described these cases to illustrate how women, though not themselves working abroad, can employ migration to their own ends. These women do not openly call into question, or revolt against, their position within the household's hierarchy, but they remain integrated in the village society's normative system and conform to it. They have realized the chances the migration of their husbands might offer, and for them it is an informal strategy (see Beck and Keddie 1978) to enhance their position. Rather than lamenting about suppressive structures and their inability to overcome them they find out potentials inherent in these structures and use them for their own interests, similar to what Macleod (1990) in her book about the new veiling of Cairene middle-class women has called their 'accommodating protest'.

CONCLUSIONS

If the relevance of international labour migration for the peasant subsistence-producing household has found the interest of scholarly research at all, this has mostly been in some kind of structural terms where the effects or consequences of the macro structure upon the latter have been focused on, presenting the peasantry as the inevitable losers, as only able to react. International labour migration is often seen to have negative effects and finally to lead to the dissolution of the peasant household. Yet, as the assumption of the household's dissolution in the face of market integration departs from questionable theoretical premises, migration also must not inevitably lead to the former's vanishing. The empirical evidence given in this chapter clearly points to the fact. Of course, the peripherization of large parts of the Egyptian agriculture has been a constant feature since the beginning of the last century, but the final breakdown of the household economy has never taken place. Very much to the contrary. In the Egyptian countryside small subsistence-producing peasant households have not disappeared even after two decades of mass migration. Two decades of massive out-migration have thus not been characterized by 'primitive accumulation' (Meillassoux 1976: II, Ch. 3). Peasant households in Kafr al-ʿIshra, on the whole, do not 'waste' remittances on purely consumptive purposes. Rather, strategies and investment are clearly linked to the aim of (re)consolidating subsistence production to the amelioration of the very basic conditions of the reproduction of life and hence of labour power. Yet, this persistence in the age of massive out-migration has been characterized by what I have previously termed the 'stability of instability'.

This chapter has shown that the majority of small peasant households are not able to reach a permanent and decisive re-organization of their household economy: remittances are not enough to purchase a sufficient amount of land to secure production for consumption, thus enabling them to reach at least relative autonomy from the market. Neither can they establish a small 'off-farm enterprise' – for example, by buying a taxi which earns them enough cash to guarantee the household's survival. What is possible instead is investment into the improvement of living conditions: scarce living space can be enlarged by building a new house or additional rooms, a cow can be purchased, clothes, nutrition and medical treatment can be bought, a marriage can be arranged. When remittances run out 'everybody is where he started from' as one village woman from Kafr al-ʿIshra stated. What was achieved then was the improvement of the very basis of subsistence production. The temporary full-scale proletarianization of part of the household's labour power on foreign working sites is one strategy within this overall scheme of households inventing an ever-new variety of strategies to secure the satisfaction of their basic needs within the capitalist market economy. But, given only mediocre returns, the 'stability of instability' is perpetuated.

Here, it is also the temporary character of migration which leads to the opposite of the postulated destruction of subsistence production. Migrants' stays abroad do not turn out to be for good, as is the case, for example, with the 'second or third generation' Turks in Germany. Egyptian peasant migrants do return and thus migratory wage labour contributes to the persistence of the small peasant household in a time of ever-increasing capitalist market integration.

Peasants, then, are not passive, merely reacting to macro structures, but they perceive of migration as a chance to reach their aims. This, then, brings us to a concept of the peasantry as actors purposefully developing survival strategies and translating them into action within a given structure. From this point of view international labour migration no longer appears as some kind of structural force inasmuch as it is a medium to achieve one's aims. International migration is seen by the social actor as a 'resource', to borrow a term of Giddens (1977: 129–34). Migration is viewed as an opportunity, a chance, to enhance the individual household's capacity to function as a unit of production and consumption, and, inseparable from that, to work for an enhancement of status. The migrant households try their best to make optimal use of this macro-structurally-given opportunity. In this sense Thurley and Wood have noted that

> the idea of strategy does seem to imply an external force or forces which one must anticipate and try to come to terms with; strategic thinking arises from the need to cope with such pressures, not because they can be ignored. Lack of autonomy and strategic thinking are not then mutually exclusive.
>
> Thurley and Wood 1983: 2; cited in Crow 1989: 20)

In the light of the preceding histories and cases from Kafr al-ʿIshra this concept has gained multifarious evidence. Strategic conduct does not a priori emerge from positions of power, autonomy or freedom of action – rather the contrary may often be the case. Constraint does not make impossible the pursuit of one's goals. And, it only appears to be natural that in such situations of coerciveness the creative drawing on whatever may be turned into a resource is of all the more vital importance. In the 1970s and 1980s, in a situation of the small peasantry's deteriorating terms of agricultural production and with growing monetization putting on them ever-increasing demands for cash, international labour migration clearly occupied a paramount place in the strategic planning of peasant households. Its pre-eminence results from the expected monetary returns, as well as from the place it occupies within the orientation and expectations of the household.

While securing the subsistence base clearly emerges as the most important impetus behind migration of small peasant households, this is not the exclusive aim: other, status-related, goals are followed simultaneously. This is all the more the case as there are of course a number of returnees who have accumulated substantial wealth and climbed up the socioeconomic ladder. In this sense migration is viewed not only as the sole realistic chance which might lead to a living above the subsistence line, but also as the only way to an enhanced status. Both aims are inseparably linked, as one aspect of the use-value of a product is always its function as a marker of status and of a certain life-style. Any act – spending remittances, marrying, dressing – has a symbolic dimension which can be used on the village community's cultural battlefield.[25] Nothing is wasted.

Migration, then, is the temporary full-scale proletarianization of the household's male labour force on the international wage labour market, and its aim is the economic stabilization of the household as a unit of (use-value) production and consumption, as well as enhancement of status.

This brings us to the concept of 'the household', and migration as a household strategy with economic contents. The empirical evidence presented in this chapter has shown that households are far from being monolithic blocs, as one might perhaps deduce from the conceptualization introduced by Wood (1981, 1982). A more appropriate concept would imply that households form collectives only inasmuch as all their members subordinate their individual interests to a common goal. This, very often, though not always, is the result of a struggle over interests and legitimacy, the outcome of a power imbalance where the paterfamilias is able to impose his will upon all the other members. This, on the whole, is the situation in the cases I have discussed in this chapter. In the extended household the senior couple have a primary interest in the household's consolidation of production and in status enhancement, and conflicts of interest are often 'solved' in their favour (see also Pahl 1985). In this sense Friedman has remarked that

> Mutual dependence does not guarantee lack of internal conflicts or basic differences of interests. In fact, it may well be that historical instances of family solidarity are what must be explained, rather than assumed.
>
> (Friedman: 1984: 41)

Yet reality is more complicated still. Households consist of distinct individuals who in one way or another are forced to subordinate their personal goals to those pursued by others who are more powerful in the household's hierarchy. But equating the household's goals and strategies with those of the most powerful in rank may also remain beside the point as it deprives us of an understanding of the 'fitting together' of distinct particular interests. In this sense, I have shown how sons migrate, because this is what they are expected to do at a certain point in their life-cycle. They migrate in order to fulfil the precondition of marriage, which in turn is the precondition for the household's future potential for procreation, the reproduction of its stock of labour power. But at the same time, the sons more or less secretly follow the aim of establishing their own household in order to escape from the authority of, and subordination to, those of higher rank. Very similar aims may be followed by other household members occupying a similar structural position in the household, who, however, do not migrate – these are the daughters-in-law.

7

ORGANIZING THE HOUSEHOLD'S SURVIVAL DURING THE MIGRANT'S ABSENCE

It has been my argument that the migrant household's continuing existence and its functioning are of crucial importance within the migratory process. If the household ceased to exist, if migration ended in 'primitive accumulation', then the simultaneous production of cash-crops could no longer take place. Furthermore, the production and reproduction costs of the labour force could also no longer be transferred to the subsistence-producing context and the reintegration of migrants would no longer be possible; peasant migrant flows would be endangered. Of course, the migrant and his entire household have an interest in the 'independent' survival of the unit and its functioning during the period of migration. For one thing, the migrant wants to return home and resume his former position in the household. After all, it has been the very aim of migration to reconsolidate or to establish a household. On the other hand migration would probably be impossible were there not the household members staying behind with their networks of cooperation. The migrant household, those in the village exploiting the household's labour force, the government and international markets, all have an interest in the maintenance of the household as a functioning unit of production and consumption.

The immediate consequences of migration are obvious: with the migrant's departure the household at once experiences a considerable loss of labour power. But the absent person not only worked, most probably he also had responsibilities and exercised authority. It is thus the remaining members' most immediate concern to reorganize the household's fabric and to redistribute tasks and roles. Most of these households are of nuclear structure according to our data. The migrant sample shows that 63 per cent of the migrants come from nuclear households and that 62 per cent are also household heads.

The new situation contains particular problems for these nuclear households as here there is usually only one person who can take over a migrant's roles and tasks – the migrant's wife.

How do these women remaining behind fill the gap the migrants have left? How does the household manage to stay alive during the migrant member's absence, and how is production reorganized? These are the main questions I want to answer in this chapter. As most migrants come from peasant households I shall also focus on the implications their absence has for agricultural production. Before turning to the situation arising with the migrant's departure, I want to show that a functioning household is already a precondition for the very departure of the migrant – in other words: the migrant most probably could not even leave were there not a functioning household to finance his travel abroad.

FINANCING TRAVEL EXPENSES

The preparation of the travel, the procuring of the necessary papers, permits, tickets and possibly a contract, requires a considerable amount of cash without which work abroad could not be realized. According to the data of the migrant sample, migrants from Kafr al-ʿIshra had spent a mean of £E375 for the financing of their travel abroad. The magnitude of this sum can only be properly understood if it is correlated to the household's monetary earnings. At the time of field research this sum in fact amounted to a government employee's earnings of about 8 months or a share-cropping peasant household's monetary return of 1 *faddan* cultivated with cotton. But often travel expenses are even higher. There are migrants who indicated that their trip to Saudi Arabia had cost them £E500, and even up to £E900. The reason is that migrants leaving for Saudi Arabia must also pay for the procurement of a contract in addition to the travel expenses. While getting a contract is always a matter of access, of connections and participation in networks, the possibilities of success are definitely dependent on available funds: the higher the travel expenses one can afford, the better the chance one has to send a member to one of the rich and promising Gulf states. But in Kafr al-ʿIshra most households simply are not able to raise such high amounts of cash and, even if they could, would prefer to spend them on other things, villagers commented.

How is this amount of cash provided? As migration is the project of the household as a unit, in which all members, if perhaps for different reasons participate, it is no surprise that they also join in the practical realization and organization of this project. The cooperation starts with the financing of travel expenses. Migrants could not fund the travel out of their personal and individual savings as they rarely possess them. Money, if not secretly held back as in the case of Helwe (see Chapter 6), is usually an asset of the whole household and under the control of the household head who reserves to him/herself the right to dispose of it. As households are supposed to function as income pooling units,[1] financing travel expenses becomes the obligation of the household unit and of those household members who have personal assets. The significance of getting hold of the necessary cash to realize migration only becomes apparent if one is aware of the fact that these small peasant households generally lack cash, and that the sums at their disposal form a most important reserve for unforeseen events. And so, the entire household loses this meagre base of security through investing in the strategy of migration. This is evident, as 48 per cent of the respondents indicated that at least part of the travel expenses had been covered by savings (see Table 7.1).

Table 7.1 Financing of travel expenses (first migration)

Cash source	*Number (%)*
Savings	48
Borrowed money	23
Sold animals	22
Sold copper and gold	14
Sold land	–

Source: Remigrant sample 1987/88.
Note: Multiple responses possible.

Another way to get hold of cash goes beyond the immediate boundaries of the household to include relatives and good friends: 23 per cent of the migrants had borrowed money by calling on the solidarity of other family members and friends. Usually the father or a brother of the wife, or the migrant's father, lend the necessary money. There are also moneylenders in the village, but these are only approached as a last possibility because they are known to offer their money for exorbitant interest rates. Instead of borrowing money from such moneylenders households prefer to convert other assets into cash: 22 per cent sold an animal for that purpose – this is generally the water buffalo. As most households have only one *gamusa*, the household, through its sale, is stripped of its sole provider of milk and dairy products which are not substituted by purchases from the market. As a consequence the household is robbed of a potential source of cash income that the selling of milk and milk products had formerly provided. With the departure of the *gamusa* a basic source of security is lost as it is always kept on the understanding that cash can be made available in times of need, for example when a household member gets seriously ill. The sale of the buffalo is of special relevance to women because it is often their property. Many of them had sold their bridal gold to buy a buffalo as a productive asset. Their consent to the sale of the animal is a precondition for migration.

This is also the situation when travel expenses have been financed by the sale of the women's gold – a source in 14 per cent of the cases. The *shabka* is usually the most important possession of a peasant woman. In the village of Kafr al-ʿIshra, women often sell their bridal gold soon after marriage with the aim of consolidating the household, be that through the purchase of a buffalo or a cow, or by the investment into housing. In this sense the financing of the travel is only one form of investment. These women endorse the household's migration project and by selling their belongings they contribute to its realization. In the two streets not a single woman still had her gold, and most of them had even sold their wedding rings.

As I have shown in the previous chapter a woman's consent to collaborate is usually granted if she herself sees advantages in the migration of a household member. One should therefore not think that selling her possessions always takes place without conflicts. Here again the position a woman occupies in the household's hierarchy defines whether she will willingly turn her belongings into cash or whether this is the outcome of a power struggle. In fact a woman's assets are often only nominally hers, as other household members take it as a matter of course that everything has to be shared. The extended household of Shahhat (see Chapter 5) is such a case. When the senior couple wanted to buy a water pump they took it for granted that their young daughter-in-law would sell her gold to make the purchase possible. She categorically refused and eventually got *ghadbana* returning to her father's household. Her mother-in-law was desperate. Not only had she lost the only other female worker in the household, but was also extremely annoyed because her daughter-in-law refused to contribute anything to the household's economic progress while she proceeded from the understanding that they were one family and must all cooperate for the household's better future, especially as they were poor. What made the matter worse for her was that they are Christian and so she could not even urge her son to divorce his wife, as there is no divorce for Copts.

Due to the general scarcity of resources, the procuring of the necessary cash always includes a risk and puts the household in a potentially threatening situation. The migrants' stories about their experiences abroad presented in Chapter 5, and their mentioning how they would collect money for a newly arrived friend, are only another indication of the dramatic scarcity of funds the migrant household is confronted with when realizing the migration project. Savings and productive assets are lost and, in addition to that, with the migrant's departure the stock of available labour power has considerably declined. Still, nobody knows if these funds are returned, let alone multiplied.

COPING WITH THE MIGRANT'S ABSENCE. FILLING IN THE GAP

The absence of a household member will be felt least if the migrant is still unmarried and living in his father's household, or if he is a member of an extended household without being himself the head of the family. In these cases his tasks and obligations are still relatively restricted and there are usually a number of other members who can take over his obligations. A married migrant living in an extended household also seems to face relatively few problems as here the protection and the care of his wife and children is the duty of the household head: his father. At least this is so in the cases where women keep on friendly terms with their in-laws.

It is the nuclear household which faces most problems. With the departure of the migrant its potential of available labour force has become more or less reduced to only one person, that of the migrant's wife. Though she is integrated into the networks of the wider family and the neighbourhood, and can count upon their help, she still must adjust to a totally new situation. What often makes the tasks of these women particularly difficult is the limited base of reproduction. As migration is usually performed in the early stages of the household's domestic cycle its productive assets are rather limited: resources (land or animals) are restricted if available at all, and because the children are very young the labour force is comprised of just the couple. Many of these households are therefore dependent to a high degree on the sale of their labour power. This is no easy task for women, because a sufficient demand for labour only exists during the agricultural peak seasons. Off-farm employment, scarce as it is, is, moreover, usually for men. This situation causes the remaining women great stress and burdens them with a responsibility not easy to shoulder. Women in such a situation are dependent on receiving regular remittances, but in fact only 20 per cent of the migrants in Kafr al-ʿIshra are indicated to have sent back money to their wives while abroad. Asked for her feelings, Amina (household 4), whose husband was in Iraq for the third time, expressed her sentiments in the following words:

> I would not have any problems if he sent me money for the maintenance of the household and the upbringing of our children! I have never been responsible for financial affairs. But now, when my husband does not send money, it is me who has to look for other sources to get cash. I try to work in the fields, but sometimes when there is no labour I must turn to other households and borrow money. My husband then has to pay back the sums when he returns home.

Laying claim to the paternal household's solidarity: the migrant family's reintegration into the senior household

With the departure of their husbands, women face a totally new situation which is not only due to the problematic monetary situation. They have to assume the position of the household head, which is concomitant to an additional workload, and also the responsibility to manage household affairs. To evade this situation and its difficulties the migrant's wife and children sometimes reintegrate into the paternal household for the period of migration. This may be the household of her parents-in-law, but some women also join their own parents. It is especially the recently established nuclear families which take recourse to this option. However, women with school-age children and land or animals feel that they have too many responsibilities binding them to their own household to be able to go to the senior household. It is therefore usually newly established, economically weak households which opt for reintegration.

Sahar's case

Sahar and her husband (household 7) do not own agricultural soil nor have they rented any. Before migration both used to work as day labourers and in the fields of their parents. When Sahar's husband departed they had only recently moved into their own separate household. Sahar immediately chose to go to her father's household, taking her children and her TV set. As one of his family members she worked in the fields of her father and helped her mother at home. She did not need any money from her husband except for occasional ice cream and sweets for the children. As Sahar's own household virtually had no access to any means of production she could only have resorted to agricultural daily labour – which is a rather risky option as I have already mentioned. Engaging in wage labour would have meant leaving her small children unprotected for the time of her absence; usually neighbours or family members keep an eye on toddlers and babies, but in agricultural peak seasons this is more difficult to arrange as everybody works in the fields. Given the lack of wage earnings, and an extremely strained relationship with her in-laws, she probably would have been dependent on her father's donations anyway. It is therefore only natural that she opted to join her father's household which is not rich but nevertheless provides enough to ensure that nobody goes to bed with an empty stomach.

Sahar did not come as a guest. As a temporary member of her father's household she took over her share in duties side by side with the other members of her family, and she also occasionally worked as a daily wage labourer. In return she had the right to participate in consumption and to claim her father's protection. Going to her father was evidently the more convenient choice, and as she is his daughter he could never have refused her and his grandchildren. Sahar did not see staying alone in her own household as a matter of 'self realization' or as a means of achieving an enhanced status as household head. On the contrary, what counted for her besides her own security and that of her children was not to feel lonely in her own 'empty' house, as she said. Her family is where she belongs – and as a daugher she was not considered to be a burden. Without hesitation she decided to submit to her father's authority. She didn't see this as a loss of the personal freedom she could have enjoyed during her husband's absence, but rather as laying claim to the undeniable right of protection and care for herself and her children.

That reintegration is one possible solution preferred by the involved women is clearly expressed in the answers of other women when asked about their problems during the absence of their husbands. Many of them related the fact that they had no difficulties on their return to an extended family. One remigrant's wife said: 'I did not have any problems because I was living in the middle of my husband's family and his parents took over all responsibilities'. Another remarked: 'I am for a second migration and not against it, to secure the children's future and because during his absence I did not carry any responsibility, because I have my father.' Still another woman said:

> I was living with the parents of my husband, and his two brothers. It is the men who are responsible for the solution of problems. That is why his father is responsible for everything. He takes the place of my husband. He sent the money to his father who spent it on the family and the household.

Migration, women and agriculture

An issue of great importance during migrants' absence is the maintenance of agricultural production. Agriculture, as the peasant household's most vital source of income, is an asset too important to be neglected; no loss of output can be risked. As women are the only grown-up family members left in nuclear migrant households, agriculture has become among their main responsibilities. They cannot usually count on their children very much as these are still too young to be of any substantial assistance. The women's taking over of field work previously performed by the men before their departure has therefore also contributed to the 'feminization of agriculture' (Abaza 1987, 1988) observed in the Egyptian Delta, another reason for which is mechanization (see also Sullivan 1981). For these women agricultural labour is not a qualitative new task as they have always been working in the fields, but in the new situation arising as a consequence of their men's departure they have to give an increasing amount of time to it. For villagers the 'feminization of agriculture' due to migration has become an aspect of everyday life:

> It is known that women in the village participate in field work. They all work in agriculture. To see a woman work in agriculture is a very natural thing. But since the men took to migrate, women have come to play a greater role in agriculture.
> (Head of the cooperative)

Another man said:

> Roles of women in agriculture are no longer the same as in the 1960s when women restricted themselves to the house and to bringing up children. Since the inhabitants of the village are leaving for the Arab States [i.e. the Gulf States] the woman carries the full responsibility for the house, the field, and she has to look after the family affairs inside and outside the house. She also is responsible for anything concerning agriculture – for example, contacts with the cooperative, the bank, or with the daily wage labourers, and among the most important characteristics is that she works in the field.
> (Peasant, aged 50 years)

In the two streets of *harat al-ʿabid* the increasing involvement of women is also obvious. Although all women from landholding households already worked in the fields before migration they are required to take over their husbands' chores after migration – an additional burden. The term 'feminization of agriculture' does not, therefore, indicate an increasing participation of women who had formerly not done agricultural work as more or less all of them had been working in the fields prior to migration. What it does indicate is that they have to spend more time on these jobs. The relative decrease of female participation after the return is then due to the men's resuming their tasks rather than to the women ceasing to participate in field work. In the two streets all remigrants resumed their previous tasks in agriculture but all the women also continued to do field work. This observation is confirmed by statistical data for the whole community: of all migrants who had indicated 'peasant' (n = 31) as their main job prior to migration, 29 (93.5 per cent) still worked as peasants after their return.[2] Underlining my argument (but disappointing enough for the general outcome of work abroad) is the fact that 100 per cent (n = 40) of the migrants who had been farm labourers before leaving the village still worked as such after their return.[3] Statistical data[4] about the involvement of women in agricultural production also confirms the continuity of women's work in the field. Female participation in agriculture rose from 44 cases before migration to 46 during migration and declined slightly to 41 cases after migration. However, as the degree of involvement and the intensification of women's work in the field was not a subject of the questionnaire it could not be measured and is thus not reflected in the quantitative data.

This 'natural' taking over of what their husbands had performed often implies considerable stress for the women. Women say that they feel exhausted by the tremendous workload and the responsibilities they have to shoulder when left alone. Amina is an example in this respect.

The case of Amina (household 4)

During the cotton harvest Amina worked in the fields from about seven in the morning until late afternoon. She could not pick all the cotton on her own and hired *anfar* ('daily labourers'). As a meal and tea are included in the salary she had to prepare a lunch after returning from the fields. This meant she had to cook and to bake bread. She stayed in front of the oven until long after sunset. Baking usually requires at least three women but as her neighbours were all in a similar situation she could not count on their help. She was obliged to involve her three children, all below primary-school age. After finishing the baking she still had to hire *anfar*, which meant that she had to walk through her quarter in order to ask around who was ready to work with her. Her stress was made greater by the fact that during daytime she was forced to leave her children without company. In fact, in the days of the cotton harvest the alleys of the village were deserted except for small children in the care of their elder siblings – all the others were busy in the fields. It was as a consequence of this situation that a child had fallen into the canal and drowned. Understandably, Amina was always very scared when leaving them alone.

One way to cope with the rising workload is to call on the solidarity of women in a similar situation. As many women are also married to migrants, or know the problems from their own recent past, and also because they are friends and neighbours, they assist each other whenever possible. This practice is called *muzamala* or *zamal*, 'cooperation'.[5] According to my observations it is most often practised in agriculture, but any other work may be done in cooperation. Other examples are construction of houses

Plate 7.1 Having a rest before moving to the next rice field.

or baking bread. *Muzamala* is a non-monetarized form of production whereby the labour force from different households is pooled to get a particular task done. Reciprocity is secured by the benefiting household's obligation to provide an equal amount of labour force at a later time, or, if this should be impossible for some reason, to pay the corresponding wage. *Zamal* is therefore practised when households are in a similar situation and share similar characteristics. All of them are relatively poor and lacking in money, and therefore must try to avoid cash expenditure. All these characteristics make *muzamala* a very suitable practice. In the major agricultural seasons nearly every woman living in the two streets joined in to help, this cooperation taking the form of finishing the plot of one woman first and then moving on to the field of the next household. Sahar (household 7) enjoys this kind of work: 'I like to cooperate in the fields, because we always have much fun, we laugh and sing together and have a nice time. It is better than to stay at home alone.'

Muzamala, as a 'traditional' form of production, has not been recently introduced in the wake of mass migration nor is it a surviving relic from the past, but its increasing practise is definitely a response to migration. It is itself an aspect of modernity, a most suitable form of production within the overall mode of capitalist production. It is a strategy adapted to the requirements of the women staying behind: 'Labour migration has not caused problems in agriculture, and this has a very simple reason: villagers follow *az-zamal*, that means they help each other, for example in the cotton harvest' (head of the cooperative).

'CLEVER' WOMEN'S PARTICIPATION IN THE HOUSEHOLD'S MIGRATION PROJECT

I have mentioned before the image of the ideal peasant woman, the 'clever woman'. Women conforming to this ideal are esteemed and appreciated for their great capability and the energy they invest in order to 'make something out of nothing', in this way contributing to a gradual amelioration of the household's subsistence basis. The examples I have already cited demonstrate how women manage to keep the household and its economic activities alive during a member's absence and show that these characteristics of the 'clever woman' are in fact a common phenomenon – a capacity women must inevitably develop to cope with this difficult situation.

To emphasize this image of the 'clever woman' I will show how women, rather than restricting themselves to complying with the basic tasks needed to sustain the household, actively make their own contributions to the household's migration project. As women have a stake in migration and have adopted it as their strategy too, they also share in the practical realization of migration. While it is the husband's job to work abroad, the wife tries to develop additional sources of income and new subsistence activities. Once the decision has been taken by women to stay in their own household and resume the functions of the household head they develop great energy to cope with their new tasks. Even if they know that there is always a male relative to turn to if they need help it soon becomes a matter of pride to prove that they are capable of replacing the husband, that they are in fact not only able to keep the household alive but also to make their own contributions to its long-term prosperity. As there is no single source of revenue they can concentrate on, their strategies are a combination of many small measures which only count if applied together. In addition, they try to use remittances for investment rather than consumption purposes. While women want the security of the availability of remittances whenever they need them, they do everything not to touch them during their husbands' absence.

The cases of Fauziyya and Nadia

Fauziyya (household 20) was about 25 years old when her husband went to work in Saudi Arabia. This was some ten years ago. At that time they were still living with a brother but had already been farming their land separately. Fauziyya took over the responsibility to cultivate the land operated in the legal share-cropping system. She had made it her principle to run the household as far as possible without touching her husband's remittances. She worked on the soil and hired labourers only during peak seasons. She also worked as a day labourer. She arranged a *shirk*[6] relationship and brought up a calf. She found another small source of income by growing some bushes of *bamia* (okra), hidden between the cotton rows, drying its fruit and selling it to the village households in winter. She also collected and dried *mlukhiyya*, a wild, spinach-like herb, as women in the village usually do. She thus only very rarely buys vegetables from the market. Eventually she went to the capital of the governorate to get an identity card so that she could open a bank account to receive her husband's remittances. As she wanted to move out of the brother's house as soon as possible, she entered into negotiations with the owner of a plot of building land in the two streets in *harat al-ʿabid*. She told her husband to send her £E700 as a payment in advance until he could come to make the contract.

Nadia (household 19) also took over the initiative when her husband was abroad. In order not to touch remittances she did daily wage labour in the village fields. She eventually managed to buy a calf. She then told her husband of her decision to rent half a *faddan* to grow clover as animal-fodder and asked him to send her the necessary money. She started to cultivate the plot on her own. She said: 'This was not easy for me and I had lots of problems. And it was difficult for me to solve them without my husband. But I really wanted to manage on my own.'

Fauziyya and Nadia are not exceptional in the activities they develop to more than 'fill the gap' created by the household's migratory project. More or less all of these women engage in similar activities, and they do so not only during the time of migration. One should also beware of romanticizing this ideal of the clever and energetic peasant woman, as the meagre resources available to them limit their efforts dramatically. Often this deliberate engaging in additional income-earning or cost-reducing practices is dictated by mere necessity. As the household's economy is founded on a very weak base women simply do not have other alternatives. This is especially the case for those women who do not receive money from their husbands working abroad. These women are not in a situation where they can decide whether to use remittances for consumptive or investment purposes.

CONCLUSION

I have shown in the previous chapter how the subsistence-producing household as a unit of production and consumption is stabilized by its members who make migration a strategy to reach this aim. In this chapter I have dealt with another vital aspect in this respect. Household members remaining behind, and especially women, have to shoulder the task of filling the gap the migrant has left in the household's fabric and thereby ensure that the unit keeps on functioning in times of the absence of a crucial part of its labour force. From a structural point of view one may therefore argue that it is not only the migrant who is exploited but the whole network he orginates from. In this sense Meillassoux has maintained that

> because the work force is the social product of the community, the exploitation of a single member, as long as it is not detached from the community, is concomitant to the exploitation of all the other members. Exploitation does not only concern the single worker, but also and above all the cell he belongs to.
>
> (Meillassoux 1976: 129; my translation)

In this context attention immediately has to focus on the women remaining behind, as migrants often come from nuclear households where they had occupied the position of the household heads. Here, women, as temporary household heads, clearly emerge as a pool of surplus labour which is drawn upon during the phase of migration. Most studies dealing with the issue of women affected by migration adopt a less explicit standpoint than Meillassoux does. Here the focus usually is on the 'impacts' or 'effects' of migration on these women.[7] Such 'effects' or 'impacts' may be negative or positive – but they always implicitly depart from a concept where women appear as reactive. Here one is concerned predominantly with the repercussions, impingements, impressions or results migration has for the women. Such a procedure tends to obscure the fact that women have their own interests and try to realize them, and in this endeavour may actively use structures for their own aims. They are not confronted with the results of 'impacts' and 'effects' inasmuch as they try to put structures to profitable account or turn them into a resource. That this is in fact the case has been already demonstrated in the previous chapter when showing that migration can also be a strategy of non-migrating women. And if I am here developing this argument further, I again depart from the fact that women occupy a certain distinct position in the household's hierarchy which is essential in the defining of her interests. Households are certainly marked by internal conflicts, but the primary identity of all their members remains within the family (see also, Tucker 1986: 52 and Rugh 1985). Thus Hélie-Lucas concludes a book review:

> All abuse of women through male authority notwithstanding all the interviews speak of the women's very strong love and attachment to the family. The ultimate punishment a woman can be submitted to is to be severed from her own family.
>
> The family embodies the values all the women stick to and which they consider as the central expression of their culture and identity: The family is part of their feminism and an element they miss in the 'western' feminism.
>
> (Hélie-Lucas 1989: 238)

The individual woman within the Egyptian peasant household conceives of her destiny as closely linked to the fate of this unit. It is still virtually unthinkable that women – or men – should try to establish a separate and autonomous existence for themselves. As the status of the household reflects on that of the women in it, adoption of migration projects by the women enhance their individual status. If the migrating husbands are successful abroad, then the women staying behind will also ultimately profit from the spending of remittances. It is with this hope in mind that they take on additional tasks, for this is their effort in the attempt to make migration a success. Sukkary-Stolba, in an article about women's roles in a land reclamation project, describes a similar attitude. Also here women are willing to take over additional work and to sell their gold because 'the new lands represented hope, better life, less crowding, and the very much sought after landowner status'

(Sukkary-Stolba 1985: 184). They do so because 'they recognized that pioneering requires extra hard work and they accepted the challenge' (Sukkary-Stolba 1985: 186).

One cannot leave unmentioned the fact that such 'pioneering' work not only consists of more hard work but that it also implies considerable physical and mental stress. Socio-psychological problems often result from the fact that the migrant women's environment is not a priori willing to accept their new status and roles. This is especially the case with older children, brought up in the conventional way where women perform the role of tender and understanding mothers and the fathers are the discipliners. Here is a typical experience of a mother:

> the absence of the father has a great impact on the children . . . . The man must be there to guide the children and to care for them. Also, they are more afraid of him than of the mother, because she is soft and emotional. Children may develop bad behaviour, especially as they get older, because of the missing guidance and care . . . . They are only disciplined if he is present. If they do not have a father around to beat them and show them how far they can go, children get used to devilish things and develop bad conduct. Children tire out their mother by behaving badly, but not their father – they are more frightened of him.

Another example is that of agricultural labourers who try to make working hours in the field pass as easily as possible. With the change from a male to a female employer they see a chance to bargain for new and more favourable working conditions. As a woman observed: 'Agricultural labourers often neglect their work if there is only a woman to supervise them instead of a man.'

Women take over their new roles without hesitation, because this is their share in the migration project within the reorganization of the household. They do so not because they see in migration and in the time of their husbands' absence a chance to achieve higher status or personal enrichment. Rather, they see migration as a strategy aiming at the long-term enhancement of the status of the household as a unit, and this also implies their own personal status. This is what the reshuffle of the household's production demands from them. While they agree to cope with the workload and any problems which arise, this is not to say that their new role is without stress. However, migration and the duties incumbent on them are expected to be of transitory character and, after the return of their husbands, they can expect to resume their previous roles, with a consequent lightening of their burdens. Here are some statements in this respect:

> I did not let anybody interfere into my new tasks during the absence of my husband. But now that he is back he is again the one responsible for the solution of the problems. I am very happy that he is back. My burden has been lightened a lot.
>
> (Remigrant's wife, 30, can read and write)

> I felt very exhausted psychologically because of the absence of my husband. There were some difficulties which need a man to solve them. For example, my daughter got ill and had to be taken to a doctor. But the presence of my parents-in-law facilitated the difficult situation a lot. I suffered during his stay abroad: I was afraid and lonely and felt the instability of myself and my family. In general I think that women struggle a lot more than the husbands do when they are working abroad. The woman is forced to carry the responsibility during his absence and this is contrary to her own wishes.
>
> (Wife of a remigrant, 27, can read and write)

> The woman carries responsibility during the time of her husband's journey but she gives this to him again the moment he is back. She feels very relaxed after his return. She then experiences the feelings of *istiqrar* [stability] and security, of peace and protection. She is then very happy because he is going to take over responsibility once again.
>
> (Wife of migrant, 22, illiterate)

It is no wonder that women see migration as a temporary project at the end of which everybody returns to his/her former roles and tasks. For women, this is not a priori a situation of increased personal freedom of action and enhanced status, but an enormous stress they want to be relieved of.

Woman can hardly cope with all these problems alone. In the fulfilment of their tasks, they are indeed dependent on the solidarity and assistance of the community. Besides their family – that is, especially their fathers and brothers – it is the help of the women of their neighbourhood that they can count upon. Most of these women are in a very similar situation or have gone through the experience previously, and others know that they might soon face the same destiny. Here it is the shared situation of poverty and deficiency which lets them take recourse to *muzamala*, to 'cooperation'. I have argued that cooperation – though non-monetary – is in fact a very modern, well-adpated form of production. Women implicitly legitimize their call for help and solidarity by referring to the *taqlid al-qariya*, the village society's system of norms and values. As they base their claim for help on *muzamala* or on the males' responsibility of protection of women, they implicitly reconstruct peasant society. Women, through laying claim to solidarity and protection, but also through their own active participation in these institutions, define what the village 'traditions' are. A similar explanation can be forwarded with regard to those women who return to paternal households for the time of their husbands' absence. Understanding reintegration as a shift in male dominance over women, from that of the husband to that of the fathers or brothers, accepts at face value the subordination of women to men. This would imply that women's subordination is a quasi-primordial or continuously guaranteed fact. But one again has to acknowledge that the social actor will exploit structures in a self-serving manner. Here it is the fact that this very authority of the males – that is, the institutionalized and publicly legitimized male prerogative of decision-making and exercise of power – also implies responsibilities and obligations women can also call upon and use in their own right. While this can be the case with reintegration into the paternal household, there are also many other ways and occasions in which women can make claim to this 'traditional' right: when they need a brother to help in the fields, when they want a male with them to negotiate with some external institution (the bank, the cooperative, etc.).

To see migration and reintegration into the extended household as an instance of men's continued control over women, and as an illustration of the latters' fate to 'never' achieve self-determination, is one understanding; seeing reintegration as a way open to women in difficult situations in order to mobilize solidarity and help by explicitly calling on the men's authority is another view. Women recognize the male domination as a resource in the sense of Giddens's:

> 'Resources' I treat as properties of structures; actors 'possess' resources in a parallel sense to that in which they 'know' rules. This is clear enough in the case, for instance, of the mobiliza-

> tion of authority rights in the context of interaction, where such rights 'belong' to the individual actor only in the sense that he can – in principle – demand and obtain certain responses from others. Authority is a structured resource that can be potentially drawn upon by actors to influence the conduct of others.
>
> (Giddens 1977: 134)

Definitely, here, it is the weak and relatively powerless who turn structures into a resource to their profit. Authority is not just a possession of the powerful, it simultaneously gives the weaker sectors of society the right to lay claim to certain obligations which go along with the possession of power and authority.

These traditional institutions, based on the community's system of mutual reciprocity, seem to me an important aspect of the 'functioning' of international labour migration within the global context. These institutions not only contribute to the realization of migration, they also secure the household's viability in times of crisis.

In order to cope with problems occurring during the absence of their husbands, many women chose the strategy of laying claim to the obligations and solidarity of relatives, thereby transgressing the boundaries of their own household. This finally brings us back to household concepts. While it has been clear from the presentation in previous chapters that households are internally diversified I have provided evidence here that household boundaries are not impermeable and static. Households must be understood as flexible units adapting to specific and changing situations. There is good reason to suggest theoretical concepts which depart from or imply the notion of process (see also, Hareven 1974, 1975).

8

AFTER THE GULF CRISIS

Revisiting Kafr Al-ʿIshra

The Gulf crisis which erupted in August 1990 led to a major break and a gradual reorganization of international migratory flows. A large part of the migrant population of Iraq, Kuwait and also Saudi Arabia was forced to leave in the initial period of the crisis between August 1990 and January 1991. According to one source

> more than 2.6 million persons were displaced. This figure includes at least 1.6 million refugees who fled Kuwait during the Iraqi occupation, as many as 700,000 Yemenites forced to leave Saudi Arabia, and as many as 400,000 foreign nationals, mostly Egyptians, who managed to leave Iraq before the allied bombing campaign began.
>
> (Hooglund 1991: 5)

The same source continued: 'more than 500,000 Asian workers fled Kuwait in the initial 10 weeks following the Iraqi invasion' and most of the 100,000 to 150,000 Asians, in Iraq, predominantly from the Indian subcontinent, also left the country (Hooglund 1991: 5). Concerning the Egyptians Hooglund gives the following estimation:

> At least 350,000 Egyptians managed to get out of Iraq by mid-January. Including the estimated 135,000 Egyptian workers who returned from Kuwait, this meant approximately half a million extra workers looking for jobs in an Egyptian economy already plagued with high levels of unemployment, and the loss of hard currency remittances estimated at over $1 billion dollars per year.
>
> (Hooglund 1991: 6–7)

In early 1992 another report, by Fergany, corroborated these figures to a great extent. According to this source 'about two million Arab expatriate workers, and their dependents, were removed from their residence . . .' (Fergany 1992: 12). Approximately 700,000 Egyptians must have returned (ibid.) and 'the most plausible estimates place the number of Egyptians in Iraq in the neighborhood of 100,000 only' (ibid.: 15). According to Fergany only the very well integrated and the very destitute have remained while all the others have returned. These remaining persons have been living in Kuwait or in Iraq for a long time, or have established close links to the indigenous society through intermarriage or through entrepreneurial connections. But according to him, those who saw no chance for reintegration also stayed. In Jordan, a worsening economic crisis, due to its political position during the war and the following international punishment which cut off aid and regional export markets, also resulted in the return of migrants. Libya, on the other hand, has once again opened its frontiers for expatriate labourers, but according to Fergany's assessment its economy cannot accommodate significant numbers of migrants (Fergany 1992: 19) to qualify as a new outlet for migrants. He also points to the still increasing preference for a non-Arab (i.e., Asian) labour force in the Gulf economies after the war (Fergany 1992: 17).

However high the number of Egyptians working in Iraq may actually have been, it is very obvious that most of them have returned. But in spite of this massive return movement and the continuing economic crisis in Egypt, no revolution has occurred. Rarely is the fate of these persons discussed in the media nor does it seem to find any public interest: there is no 'returnee question' and the migrants seem to have been forgotten. It is at this point that my thesis of the usefulness of the subsistence-producing peasant household and its capacity to reintegrate the migrant member becomes of interest again. My major argument in this respect has been that the functionality of the peasant household for globally organized labour markets is grounded in its capacity to produce and reproduce this labour force – that is, that this task is externalized in the capitalist process of production. Here, the reabsorption of the superfluous labour force becomes important. In summer 1992 I therefore went back to the village to learn what had happened to the migrants and their families. I wanted to see whether the migrants had in fact been reintegrated and whether this might explain why there was no 'migrant question' on Egypt's political agenda. It was necessary to investigate which economic activities the small peasant households in a 'post-migration phase' engage in to secure survival. Though I do not have statistical data for the whole community, my observations in the two streets I had formerly worked in, and in *harat al-ʿabid* in general, turned out to underline my arguments.

ORGANIZING REPRODUCTION AFTER THE RETURN OF THE MIGRANTS

In the two streets all the migrants have returned but one (household 16), who is still working in Jordan. Two persons (1,7) had left for Iraq after the termination of my fieldwork and both of them returned shortly before the outbreak of the war. Two other household heads (3,4) who had been in Iraq have also returned. A rare exception in the village is the fiancé of a young girl (1) who has not come back yet. After a long period of uncertainty whether he had died during allied bombardment it turned out that his employer had taken away his passport and thus hindered him from quitting his job. Nobody thinks of working in Iraq anymore. Migrants to Jordan have also come home. Thus, the household head of (8) has been back for a long time, while the migrant member of household (12) returned during my stay in the village. The latter came back because of the Jordanian intention to charge JD150 per year from each migrant, making work abroad for most of the unskilled migrant workers no longer worth while. Only a married son of household (16) who had left after my field stay does not want to return. He has mailed a cassette to his family informing them that he intends to stay and pay the sum. This development of return migration from Jordan seems of very recent date and to be closely linked to the new fees demanded, making work in this country no longer profitable for the low income groups.

However, after the outset of the Gulf crisis, Jordan, as did Libya, seems for a while to have constituted an alternative for the villagers. While not the case in the two streets, in *harat al-ʿabid* in general there are some households which have sent a member for work to Jordan or Libya in the past. But it is generally agreed upon in the village that work in Libya is not rewarding, for not only is it difficult to find a job but also workers are largely dependent on daily wage labour and these wages are low. There are also some migrants who left for Kuwait. All of them seem to participate in well-established networks as they have generally got hold of a contract through the help of a relative already working in there.

Without this help prospective migrants have to buy a contract in Egypt at around £E2,000 to 3,000, which definitely is outside the financial scope of the small peasant households. While most of the remigrants stated that they would depart again if the chance arose, it is generally agreed that work abroad is no longer a viable option at the moment.

Obviously all the households in the two streets have the capacity to reintegrate the migrant members. Each has returned and resumed his specific role in the household's fabric. Often the role is identical to what had been done prior to departure but remigrants also join the new economic activities the households are now engaging in. Another obvious fact in these two streets is that no migrant household, and none of the non-migrant households for that matter, has made any substantial economic progress in the last years. Only one non-migrant household (15) which had already been comparatively well-off through landed property and a workshop is today clearly in a better financial situation. This recently achieved status is symbolized in its move to the village's 'industrial zone' at the asphalt road, where the household has built a relatively large red brick house where formerly agricultural soil had been. This house also comprises the facilities for two shops in the basement, which are not yet finished and not rented. The workshop is run by two to three labourers who, though working relatively permanently, are paid on a daily basis. One youth of the two streets (3) has found a job here. The new status as successful entrepreneur is visibly demonstrated by the husband's wearing white *gallabeyya-s*: he does not work with his labourers but has taken over the task of collecting and redistributing the water-pump tubes with his recently bought pick-up. His enhanced position in village society is further demonstrated by a new friendship with the ʿOmda with whom he spends many evenings smoking a water pipe.

As migration abroad is no longer a viable option at the moment, the question arises as to how the poor, small peasant households manage to secure survival today. Very obviously, and not surprisingly at all, engagement in agricultural production and wage labour emerge as most important. As it was before remigration, agriculture is still a main source of income and economic strategies are predominantly linked to it. As a matter of fact, no household in the two streets has been able to buy land. This is not surprising as land prices are still dramatically on the rise and in summer 1992 reached

£E30,000 to 50,000 per *faddan* according to the quality and accessibility of the soil. Without doubt the purchase of even a tiny plot of land is beyond the scope of these households. Yet, there are three households (5,7,19) which have access to more land now. The tailor household (5), which formerly had not engaged in any kind of agricultural production, now rents half a *faddan* to grow wheat and maize for home consumption. This costs about £E400 per cropping. The work is entirely performed by the women of the household – the elderly female household head and her two daughters-in-law — while the sons are still exclusively occupied with tailoring. Household (19) also rents half a *faddan* to grow subsistence crops. Only Sahar (7) has somewhat more permanently consolidated her basis of subsistence. Formerly a landless household, Sahar inherited 4 *qirat* when her father died as another victim of the asphalt road. She and her husband also bought from her brother 10 *qirat* operated in the *muzaraʿa bin-nuss* form for £E1,000.

All the remigrants have taken up field work again, be it on their own fields or as daily wage labourers. But this hasn't led to any decline in female participation in agriculture and there is definitely no female disappearance from the 'public sphere' observable. In this respect, obviously, poverty makes any discussion about the repercussions of migration and remigration on female status superfluous. Now, as then, every hand is needed to make ends meet. Farida's (12) answer to my question of how far had women's lives changed through remigration is only too obvious in this sense: 'Women always have a hard life during the migration of their husbands. But we always have to work very much, whether our husbands are here or not. Just look at our village!'

Besides agricultural production other forms of household production are practised. The household of Hagg Salim (18) shows that subsistence production is as important as it has always been and, given the loss of income from remittances, it might actually even become of greater significance.

The case of Hagg Salim's children

Hagg Salim and Aisha were both working in the fields when I met their sons at home. They related how they were looking after their home and caring for their little sister, and that they also knew how to cook in order to feed the family when their parents were away in the fields. Ibrahim, the eldest, had bought from a school friend a pair of rabbits for £E27. This amount of cash was the savings of the four brothers from last cotton harvest:

> *Anfar* do not work very thoroughly in the harvest and therefore much remains on the cotton plant after work is finished. Then the children go and collect what is left. Every child gathers as much as possible and we sell the cotton to Samir [the grain trader of the village]. He pays £E1.5 per kilo. Each of us was able to earn £E3 to 5 per day last year. From this money we bought the rabbits. Our parents did not give us money. The rabbits very quickly bred and we soon had 30 of them. We sold three on the market and ate four. From the returns from the three rabbits we bought two geese. They are also ours and we are responsible for their fodder. When they grow up we want to sell them and to buy two goats, and then a pair of sheep, and then a calf. But when will this be? This certainly will take a very, very long time . . . . Then, a cat came and every night took away one rabbit, until it had killed ten. But we later detected this and killed it with a poisoned fish. The rabbits have also been ill and we had to buy some medicine for them. We bought the medicine from a school friend where it is much cheaper than in the pharmacy. Now, in summer, they do not breed because of the heat. We are responsible for them and we do everything for them; our parents have nothing to do with them. But my mother goes to the market and sells them so that the traders will not cheat us.

Then, the children showed me the rabbits' and the geese's living quarters inside the house, and their eyes shine when they enthusiastically start to tell me again of how they earned money from collecting what was left on the cotton fields by the *anfar*. These children obviously also conform to the ideal of the 'clever woman' described before, and they are very proud of making their contribution to the household's subsistence production. In a time of continuing severe economic crisis even these small measures, which include their catching fish in the canals, are of importance and are seen as a chance to secure survival. Observing Hagg Salim's and Aisha's household over a period of four years, and remembering its past as a landless household, several stays abroad and its investing of remittances in land and animals, plus the couple's educating of their children to make contributions to the household's subsistence production, one can clearly see their success, modest as it may be.

Yet however important the recourse to such 'traditional' forms of production may be, cash income in a time of increasing monetization of the village economy is an urgent necessity. And thus we find Hagg Salim and Aisha not only working on their own fields and encouraging their children to make their own contributions, but also engaging in various forms of wage labour. The easiest way to get hold of cash surely remains agricultural wage labour. Yet, contrary to land prices, wages have not significantly changed over the last years. Still, between £E2–5 at maximum are earned according to the working task. But other strategies are also discernible. Thus, I found that Sahar (7) had turned into a very active petty trader selling vegetables, white bread and fuel at the front door of her house. How desperate and ill-rewarded this strategy actually is may be understood when looking at her income. She told me that she earns approximately £E3, and at maximum £E5 per day. Here is just one example of her meagre profit margin: in town five round flat loaves of bread are sold for £E0.25, she sells four for the same price. I generally got the impression that there are more vegetable selling enterprises in the village now. Besides field work, this is almost the only other option open to the women as they do not perform wage labour outside the village. In the two streets there is only one woman (1) who every second day helps in the household of a large landowning family and thereby earns some money.

Nearly all the households in the two streets, in addition to agricultural wage labour, now set at least one male member free for daily wage labour outside the village. The easiest way to find work is to turn to the district town's 'slave market' where employers and employees meet at about seven in the mornings. Mostly, it is jobs in construction or loading and unloading trucks which can be found, work which is remunerated with £E5 and where only the young and healthy have a chance. Most households (except for 1,5,9,15,17) send a member to the 'slave market'. Nadia's husband (19) sometimes goes to work for a contractor in Cairo. He is a plasterer and therefore a skilled worker, which makes finding a job slightly easier. He usually stays for some weeks or a month in Cairo, but he also works in the surroundings of the village from time to time.

Another very obvious and definitely the most striking consequence of remigration is the reappearance of *tarahil* labour in the village — though this is not (yet) the case in the two streets. *Tarahil* labour is clearly of increasing significance in the village. This is all the more important as I did not come across one single case of *tarahil* when I was in the village in 1988. Arif, the son of Shahhat (see Chapters 5 and 7), has already done this form of labour for the fifth time. There is a contractor from a neighbouring village who picks the men up on a truck and drives them to Port Said where they work for a month digging ditches for telephone lines. As in former days they take along their nutrition, usually consisting of bread, onions, sugar, tea and cereals, and they get new rations sent from the household so that they are virtually independent from the market. *Tarahil* labour usually consists of a working period of approximately 25 to 30 days, after which the labourers return home. After a short stay in the village they leave again. The labourers I met were paid £E112 per month while the contractor took at least £E8. Arif told me that he had departed with twenty other men from the village, but there are also other contractors coming to the village and recruiting labour for Port Said. There are also villagers doing contract labour in the new desert towns, in hotel constructions in South Sinai or in Cairo, or who work in the oil industry. Most villagers agreed that *tarahil* labour is now a common phenomenon and they attributed its rising importance to the consequences of remigration. In the time of massive migration to the oil-producing countries nobody was inclined to do this kind of job because work abroad was far more rewarding. As this is no longer the case *tarahil* labour turns out to be a new viable option. While returns are meagre, this form of internal migration has the advantage of being adjustable to the needs of agriculture. Villagers practised it when their labour force was not needed in the fields.

The rising importance of *tarahil* labour, as well as of the 'slave market' in the district town, as a source of cash income is definitely a consequence of the restructuring of international migratory flows. Villagers have found in them new opportunities to get hold of much-needed cash. While the state had, in a sense, profited from its loyalty with the allied forces during the war and found its position financially rewarded,[1] the poor peasant households were definitely the losers. Not only did the migrants lose their jobs, but many of them are still waiting for the return of their assets which were frozen when the conflict started. No agency has helped them to overcome their hardships. Yet, all of them were reintegrated into their households. It is thus this institution that had to cope with the difficulties arising from the change in job opportunities for migrants.

SYMBOLIC STRUGGLES OVER REINTEGRATION AND THE SHIFTING MEANING OF COMMODITIES

The smooth reintegration of remigrants into the fabric of their households and the village society in general cannot be understood by looking from the economic angle only. Remigrants must also become accepted members of the community again. There has been an intense discourse observable over the meaning of remigration, the identity of the migrants, the construction of the self and the other. Villagers often expressed their doubts about the remigrants' willingness to reintegrate into the community's moral universe, and to become one of them again. In the following two quotations some of the suspicions and stereotyped ideas about the negative changes in the migrants' behaviour and identity find expression:

> Migrants walk through the village clothed in white, ironed *gallabeyya-s*. They put the *qufiyya* [head scarf] of the Arabs [i.e. the people of the Gulf states] on their shoulders. They have radios in their hands and cigarettes in their mouths. They tuck up the sleeves of their *gallabeyya-s* so that one can see their expensive watches.
>
> (Non-migrant, 32, wage labourer, can read and write)

> Migrants have learnt bad words abroad. Their children learn the bad words from them. And they also drink a lot of beer and other things which make them drunk. In past times villagers have visited each other, that does not happen any longer.
>
> (Non-migrant, 45, illiterate, wage labourer)

The 'white, ironed *gallabeyya*' is indeed foreign to the village. These are not only different in colour – as usually peasants prefer a dark material – but also in style (see Rugh 1989). While the peasant garment is a very loose-fitting *gallabeyya*, the robes of the Gulf Arabs tend to be much more form-fitting. This dress, made from a slightly transparent tissue, has been introduced in Egypt's rural areas through migration. Usually, migrants bring back the material and have it tailored here. Again, we find the white colour as a symbol expressing distance from the dirt and mud of the village: the one who wears a white and form-fitting *gallabeyya*, who walks around in the village with a cassette-recorder and smokes cigarettes (while all the others smoke the cheaper water pipe), does not have to do what most villagers have to – that is, toil in the fields. Clearly this showing-off behaviour of the migrants, their distancing themselves from the community, is frowned upon and observed with suspicion. The change in the migrants is, however, not merely superficial for they no longer conform to the villagers' moral code and even pass this defiance on to their children. They are no longer *ʿishariyyin*, and this will eventually threaten to erode the cohesion of the community.

As the show-off behaviour of migrants demonstrates the distancing of themselves from the community, so does the villagers' contempt for the migrants' deviation indicate the exclusion of the returnees from the community. For those who earned enough money abroad to secure a permanent improvement of status, or for those who have secured for themselves a permanent job abroad which makes them independent of the village economy, this stigmatization and the exclusion from the 'community of the poor' might not pose major problems. For them, the village and its community have turned into the 'rural idyll' where one spends one's holiday and leisure time. These persons do not risk much as they are no longer dependent on the village society's solidarity and its networks of reciprocity. They have marked their distance from the village-community through their demonstrative exhibition of the symbols of modernity and through their new habitus. And it is most likely that in spite of the villagers' disdain the sitting room of the successful migrant 'will never be empty of the young men who come to see him. All of them want to know whether he can provide them with a contract' (see the story of the successful migrant on p. 116).

However, as I have shown, most of the migrants from small and poor peasant households still belong to the same social stratum after their return from abroad. Their undisputed acceptance as members of the 'community of the poor' remains of great concern. They cannot risk losing their neighbour's and their family's readiness to include them in the moral obligations of reciprocity. It is a necessity for them to secure their place and their membership within the community. The question arises as to how these migrants from small peasant households see their standpoint within this discourse and how they manage to safeguard the solidarity of those left behind. It might well be precisely through the explicit stressing of their experience of the outside world that they find a possibility to reoccupy their places within their community. Their sojourn abroad, in a distant country, beyond the reach of the rest of the community, may be turned into a resource which brings them new authority. As they got to know the habits and manners, the way of life of distant peoples, they have gained experience of the outside world, and their horizon has widened. As one non-migrant from Kafr al-ʿIshra said: 'in our village, people respect each other. When a migrant returns, he is even more respected, because he has been abroad and because he has met many people and has learnt a lot from them.'

It is the demonstration of the experience of the outside world and a parallel positive identification with the village community which secures reintegration. Knowledge and identity find symbolic expression in the enumeration of the positive and negative sides of the host country's society, while simultaneously the superiority of the village community's moral system is stressed. In rather stereotyped statements migrants stress the better infrastructure in the host country (thus indirectly criticizing their own government for its inability to solve the country's urgent *azma-s* ('crises')), while at the same time blaming the foreign society's moral deviation:[2]

> With regard to religion there are very few Iraqis who pray and who believe in God. To my mind the Iraqis have no morals and one very rarely finds an employer who has morality. Especially the behaviour of women and girls is bad, they are without morality. And Iraqis drink alcohol. But Iraq is a good country. Its economic system is very good, in spite of the war.[3] I appreciated their better order. They have better facilities. Roads there are good. People comply with the laws. Bribery is virtually unheard of and is a hanging offence. The standard of living is high. There are enough goods on sale and everybody is free to buy them; nobody needs ration cards. And prices are reasonable. But the high taxes the Egyptians impose, that is too much.
>
> (Remigrant, 26, clerk)

Another example is Hagg Salim's narrative of his experiences in Saudi Arabia: 'There are many Saudis who engage in moonshine distilling. Not all of them, but many. My employer also offered me some alcohol. But I never accepted any. This is against the *sharīʿa* [i.e., the Islamic law].'

Here, the struggle over the evaluation of 'modern' and 'traditional' life-styles emerges again as an important theme. The community of the 'left behind' and the relatively unsuccessful remigrants are in agreement that modern life-styles imply moral decay, whereas village tradition upholds morality. It is perhaps no wonder that within this discourse opposing and conflicting images of the traditional peasant woman are of great significance. Here, we find the symbolic meaning of peasant women, as that of the *gemeinschaft* of the rural peasant community, as a matter of intense struggle. Here again we can observe the constant shifting of meaning. While the village and those occupied in agriculture, those working with the mud, are at one moment the backward, the traditional, the despised party, at another they are the most powerful images for one's own roots. Abaza pointed to the negative connotation of peasant women, when she wrote:

> To till the soil today is a despised activity which is not very lucrative. One can argue that this was historically always the case. Perhaps the difference is to be found in the fact that such dichotomies (rural/urban, 'civilized/non-civilized', etc.) increasingly meet gender divisions. In fact, to cultivate the land is a task suited to women who in that way can make a small contribution to cover living expenses. The 'real' wages are brought home by the males, whose work, more than ever before, lies outside the village.
>
> (Abaza 1988: 193; my translation)

Here women are the most potent symbol of stagnation and backwardness. But in our context the evaluation is again turned into the absolute positive:

> Village women are the best one can take as a bride. If you chose a stranger from outside, perhaps even from the town, you never know what you're getting. But we know our women, they are all hardworking in the fields and in the household. And they are good *sharikat hayat* [partners for lifetime], they work side by side with their husbands.
>
> And when the husband leaves for abroad, she performs a very important role: she works in the field and she works

> together with the cooperative and the bank. She faces many problems and she has to suffer very, very much during the absence of her husband. But she knows how to supervise the construction of the house, and she starts to play an important role in village society during his absence because she carries the full responsibility during the absence of her husband.
>
> (Hagg Salim)

Tucker comments on this positive connotation in the following way: 'Indeed, the idealization of women and their role may have its greatest impact in moments when the objective situation of women is changing rapidly, and the search for stability and continuity lends ideology more force' (Tucker 1986: 9).

Another example of the remigrants being in accord with the community's value system is their sharing in reciprocity. This, for example, is the case with *nuqta*, small cash gifts on major events and festivities in the life-cycle of a person or a household. *Nuqta* functions as a kind of 'social insurance' in times of need. On the occasion of a wedding Hagg Salim (household 18) gave a *nuqta* to the bridegroom's family. Asked for the meaning of this institution he answered:

> *Nuqta* belongs to the very old and important traditions of the peasants. We all assist each other in times of need. We give money to the family, be it for a wedding, a birth, a circumcision, for all the joyous events, when money is needed. This is a very old peasant tradition and we are proud of it. The receiving family writes how much it has got from each donator and will give back on a future occasion the same amount, and possibly a little more. It is like a money-box. I gave only £E5 now because my next occasion will be in about ten years time when my eldest son will marry. But if this occasion had been in the near future I probably would have given £E20. It is not important how much a family contributes because we are poor and do not have much money and because every donation is registered, but we all participate in our neighbours' and friends' joyous events.

This dependency on reciprocity takes other manifold forms. Other examples are the emphasis on *muzamala* ('cooperation', see Chapter 7), or the partial return to the mutual obligations inherent in the system of the extended household:

> Since I have established my own household, I do not have any problems any longer. And if there is a problem, it is quickly solved. At harvest time, I take a holiday from my job and I help my father with the harvest. At the first of each month I buy food for them: what I buy for myself I also buy for my parents. When occasionally I go to Sinbillawein to buy something, I buy the same for my parents. I try not to make my parents sad. If my parents have problems with their neighbours I am on their side. On Mother's Day I buy a present for my mother because I have not forgotten that she raised me. When they need me, I am at their side, and I shall not leave them until the day I die.
>
> (Migrant, 29, civil servant)

After the dream of independence from the patriarchal hegemony has been realized by the investment of remittance into a separate household, the 'rebellious' son discovers that life outside his father's protection also has its considerable disadvantages. Having realized household separation they gradually come closer to their father's households again by complying in a demonstrative way with a son's obligations towards his parents.

It is precisely by their explicit stressing of, and participation in, the community's system of reciprocity that the remigrants manage to lay claim to equal treatment in a practice developed to help the very poor. It is through this mechanism that they try to exploit to full advantage what they had (re-)defined as the moral obligations of the poor peasantry. Here, we participate in a process of constructing what Scott (1985) coined the 'remembered village', what Anderson (1983) termed the 'imagined community' or what Hobsbawm and Ranger (1983) meant when speaking of the 'invention of tradition': the peasant *gemeinschaft* is reconstructed in a time of rapidly changing overall structural conditions which confront the small peasant households with massive negative effects and impose the burden to cope with them. It is also illuminating in this respect to bring in Bourdieu's observation that

> [t]hus we see that symbolic capital, which in the form of the prestige and renown attached to a family and a name is readily convertible back into economic capital, is perhaps the *most valuable form of accumulation* in a society in which the severity of the climate . . . and the limited technical resources . . . demand collective labour. Should one see in it a disguised

> form of purchase of labour power, or a covert exaction of corvées? By all means, as long as the analysis holds together what holds together in practice, the double reality of intrinsically equivocal, ambiguous conduct. This is the pitfall awaiting all those whom a naively dualistic representation of the relationship between practice and ideology, between the 'native' economy and the 'native' representation of that economy, leads to self-mystifying demystifications: the complete reality of this appropriation of services lies in the fact that it can only take place in the disguise of the *thiwizi*, the voluntary assistance which is also a corvée and is thus a voluntary corvée and forced assistance . . .
>
> (Bourdieu 1977: 179)

Modest monetary results of work abroad force migrants to build up for themselves a reputation as good and respectable members of the community and to (re)construct the village society's system of norms and values. As there is no chance open to them to guarantee an existence outside the village context, or independent of the village society, they finally have to return and reintegrate. Any deviation from the village's moral universe and its networks of reciprocity would thus be almost concomitant to social suicide. And it cannot be overlooked that in this endeavour they ultimately join forces with the large landowners in the village, dependent like them on the village-based resources.

Coming back to Kafr al-ʿIshra at a moment when the processes of reintegration of the migrants are almost completed, symbolic struggles of the kind described above are observed to be of decreasing import. As the migrants of small peasant households have actually been able to reintegrate into the village economy and have taken up again their part in the households' task this sort of symbolic dispute has lost much of its meaning with regard to this group. However, they may have become even more acute for those who are still abroad and who have managed to retain their jobs and have proved their superiority: they can still 'walk through the village clothed in white, ironed *gallabeyya-s*' with Marlboros in their mouths.

Returning to Kafr al-ʿIshra it seems to me that the discourse about the 'right' evaluation of modernity and tradition is as virulent as ever but that the content and meaning and representation and emphasis have already changed, or have gained additional new dimensions. As the superiority of village tradition is once again undisputed, signs are conferred upon new meanings. Buying consumer durables is no longer seen to be the privilege of migrants – all the small peasant households are on fairly equal grounds in this respect. Two observations, though, struck me as quite new. One pertains to the way consumer items are got hold of today; the other relates to the ways these products are made use of. Though I do not have statistical data I met an astonishingly large number of households which buy such items in instalments. Everything may be bought this way: gas ovens, modern furniture, kitchen appliances, even jogging suits for the children. Again, these items are bought in connection with cleanliness: to the evading of the village's dirt. Kitchen cupboards are bought to ward the flies off; children want to dress in modern, colourful synthetic jogging suits, and their mothers experience that these are much easier to wash than the traditional cotton products. Cooking on a gas oven does not burn one's eyes as the straw-fired *kanun* does and does not stink like the petroleum stove.

Most important, though, and already to be found in perhaps the majority of households, is the chrome noodle machine produced in Italy. These little machines particularly aroused my interest as I had only seen them in very few households four years ago. Currently selling at around £E80, and therefore too expensive to be bought for cash, women buy them in instalments. However, village women have not developed a sudden love for Italian spaghetti (I actually never saw a woman making noodles with the machine), they use the part which produces lasagne for forming the very thin batter needed for peasant bread. By using one of these machines two women are able to cope with the tedious task of baking bread, whereas formerly three were needed as the very minimum. Without the machine women had to flatten little balls of dough into extremely thin flat cakes about 50 cm in size. This work is now done by the machine. With the rationalizing of the working process the form of bread has changed: today it is about 60 cm long and 10 cm wide. Everybody invited for a meal knows whether the household has already bought such a machine. To me, this was definitely the most striking example of a consumer durable being bought by almost everyone because it could be used for tasks definitely not intended by the producers.[4] The Italian noodle machine has virtually revolutionized home production of bread.

Plate 8.1 Baking bread the 'modern' way.

Clothing is something else I want to mention in this respect. With traditional life-styles being once more undisputed as the superior ones, wearing 'modern', urban, foreign, Islamic – in short, non-rural – clothes can convey other meanings. Different, and slightly varying, sorts of male white *gallabeyya-s* are today to be found in the village. And there is also the female (Muslim) counterpart. Such female dresses all more or less resemble or copy fashionable Cairene Islamic dresses. These are generally very colourful, made of shining synthetic material and also quite form-fitting. Different types of head-scarves are used, but these are also equally colourful and are usually adorned with little pearls and very skilfully draped around

Plate 8.2 Delivery of the *sabahhiyya*. This is an example of the merging of 'tradition' and 'modernity'. The *sabahhiyya* is a traditional custom and also a joyous occasion. The young girls are singing and clapping their hands, and they all have put on their best clothes for the occasion. All of these dresses are 'modern', yet they express different styles. These women are not dressed in the traditional black dresses put on when leaving the village, nor have they covered their heads with the usual long black *tarha-s* (tulle veils). All the women wear head-scarves in light colours. The material, size, and the way they bind them behind their heads is not typically urban or rural. Yet the second girl from the right and the one in the background have used the typical urban (Cairine) Islamic style. The latter girl, and the one on the extreme right, dresses in a kind of Cairine lower-middle-class Islamic fashion; also the dress of the second from the right is easily distinguishable as Islamic. Noticeable are the urban-style 'high'-heel shoes of the girl carrying the basket. Most of the dresses are in light colours and of synthetic fibre. These girls are not following any style connected to Muslim fundamentalism: this is Islamic, urban-style fashion as adopted by the rural women and it is only used to express the singularity of the occasion. None of the girls wears Islamic dress inside the village or in everyday life. They are walking on the main highway outside the village, and in the background we see a new red-brick house under construction.

the face. As there are no Muslim fundamentalists in the village there is also no fundamentalist clothing to be seen, whether for women or for men. But while the new male clothing does not include any

Islamic symbols, that of the women does through the head-scarves, long sleeves and skirts – these features being all that is Islamic about them as they rather express the copying of the very fashionable Cairene Islamic garment. The message is at least twofold. Often worn by young, unmarried girls they indicate a conforming to moral codes, the implication being that they will make suitable brides (which is perhaps not very important in a society reigned by face-to-face contacts). They also give expression to the wearers' knowledge of urban fashion styles. Of course, these dresses are not worn in mundane everyday life, but are reserved for exceptional occasions like a high religious holiday, a wedding, going to the governorate's capital, or delivering the *sabahhiyya*.[5] And while the materials had formerly been brought back from abroad, today they are increasingly purchased by the *tarahil* labourers on the free market of Port Said.

MAINTAINING THE SMALL PEASANTRY'S DEPENDENCY

Looking at Egypt's small peasant households and the allocation of its work-force in the two decades of massive out-migration, and then further integrating this era into the long and changeable history of market integration on a global level, the overwhelming importance of flexibility comes to my mind. Seeing out-migration to the Gulf oil economies as a bipolar movement of *homo migrans*, restricted to the years of the oil boom which have been finally brought to an end by Iraq's aggression in Kuwait and the ensuing war, seems untenable. From the beginnings of the last century onwards until the present day these small peasant households have continuously, and to an increasing degree, been subjugated to serve external interests. In quite different ways their labour force has been very flexibly put to profit by a variety of forces, be these the village-based landowners, the absentee landowners, the state, foreign industries or employers; be it through their work in the military machinery, as corvée labour, as wage labourers on the estates of landowners or on a construction site, as *tarahil* labourers or as international migrants, as share-croppers or renters. This flexibility matches the tremendous resilience with which the subsistence-oriented household organizes survival by an ever-changing combination of forms of production. Today, as in the first half of the last century, the small peasant household's 'stability of instability', within which it retains a certain degree of autonomy in the organization of subsistence production, is an essential feature. In this sense Tucker's assessment of the role of the peasant household in the first half of the last century is as viable today:

> Joint family work under the corvée and the preservation of the family structure served the interests of both the peasants and the state. The family continued to live and work together, producing for its own subsistence in the periods of relief from forced labor. The corvée impoverished the family but did not sever the bonds of shared production and consumption. It fell to the family to provide for its non-producing members, even during the absence from home. Corvée labor could be extracted by the state precisely because the family remained more or less intact, carrying on the functions of production for subsistence, provision of shelter, care for the young, old, and social regulation.
>
> (Tucker 1979: 260)

The latent endangering of the system through 'over-exploitation', leading to the verge of primitive accumulation – that is, to the final disintegration of these households – is also an integral constituent of the system which time and again can be observed. But so are the mechanisms of reconsolidation. In the last century such dangers have been the excessive recruitment of corvée labour and soldiers; the suffocating, state-ordained cropping and delivery system and its excessive demand for taxes; and the landowners' claims and share-cropping arrangements. At such critical moments 'the state revised its policies. It eliminated the most disruptive forms of corvée and military recruitment and thus lessened the drain on peasant labour' (Tucker 1979: 270–1). The Nasserist reforms of the 1950s and 1960s constituted a similar revision of policies. No really revolutionary changes in favour of the landless were carried out; rather, the system's continuation was stabilized and organized once again.

In spite of these state regulations small peasants' conditions of production gradually deteriorated again in the ensuing decades. Eventually, an occurrence on the macro-level – the oil boom – became of major importance as it provided the small peasant household with the opportunity to temporarily engage in wage labour on foreign working sites. By evolving migration strategies and investing remittances, with the aims of reconsolidating conditions of (subsistence) production and the perpetuation of its labour force, these small peasant households ultimately made a vital contribution to the continuance of the system. But it should be clear that peasant migrants to the Gulf states, and the individual household's investment of remittances, is only one rather spectacular form of the reconsolidation of subsistence production. What in the long run may turn out to be more important is what usually remains unseen and hidden 'inside the third world village': the very small but telling economic activities Hagg Salim's family engages in to secure survival – his children catching fish to avoid cash expenditure, their saving money for a pair of geese or rabbits, his wife's gathering and drying wild herbs, their sharing in the community's systems of reciprocity, their working in the fields, Hagg Salim's going to the 'slave market' and his working abroad, their eventual buying a calf, some *qirat* of land.

Over the past two centuries, no dramatic change has occurred in the overall system of production: peasants have not turned into an agricultural proletariat nor have they become self-determined landowners, nor have they left the land altogether. The 'stability of instability' is as viable as it has always been within a system which is characterized by different, often unrelated or contradictory forces, interests and constraints. We are confronted with the mechanisms of what Giddens has termed the 'duality of structure', where the small peasant producers find themselves integrated into a specific macro structure basically determined by global market forces and state agency which they turn into a resource, thereby themselves participating in its moulding. Putting relationships and dependencies in this form, one can't close one's eyes to the fact that these peasant households, over the last 200 years, have barely had a chance to put themselves outside the system or to become its winners rather than its losers. They try to use it for their own profit, which inevitably has led time and again to their own subjugation within the system of dependency.

The state's role as a safeguarder and exploiter of small peasant households – or as an institution 'not just . . . maintaining inequality but . . . producing it' and which 'transfer[s] wealth from the rural population' as Mitchell (1991: 26) puts it – became dramatically apparent again in June 1992 when the Nasserist land reforms were amended with the consent of the al-Azhar religious establishment. The new law,[6] after a transitional period of five years, will phase out old contracts between landowner and renter which were concluded for an indefinite period according to the land reform laws. The validity of such indefinite contracts had been the reason for major objections by the Shaykh al-Azhar and Mufti of Egypt (the highest Muslim legal authority in Egypt) as, in his opinion, this was against the Islamic *sharīʿa*, the Islamic law. Now, both rental agreements – that is, share-cropping and cash-rental contracts – will become invalid after the termination of the agricultural year 1996/97. The new law also regulates the termination of contracts during the transitional period. If the landowner wants to sell the land the cultivator has the right to buy it for a price both parties agree upon. If the peasant does not take over the land he is entitled to compensation equal to 40 times the land tax for each year he has left before the five transitional years are up.

Another important consequence of the new law is that rents will rise from seven times the land tax stipulated by the reform laws to twenty-two times the land tax. *Igar* will thus rise from approximately £E200 to £E630 – which is a great deal for the small cultivators to stand, especially in years of bad harvest. On the other hand this sum still remains far below the expectations of the large landowners. It came as no surprise, therefore, that during the bill's discussions in parliament some members demanded rents to be raised to 30 or 35 times the basic land tax (*al-Akhbar*, 24 June 1992) – that is to £E900–1,050. Given the fact that in the village of Kafr al-ʿIshra, renting a *faddan* for half a year costs about £E600–700 and renting a *faddan* for tomato cultivation costs £E1,000, it can be seen what impact the new contracts will have. If cultivators opt to leave the land at the beginning of the new agricultural year they would be entitled to compensation of 200 times the land tax (i.e., at maximum £E6,000); this amount, however, is far less than what they would have received under the old law's stipulations. According to these the cultivator was entitled to receive half of the land: half a *faddan* at present is worth £E15,000–25,000 in the village. High land prices, on the other hand, put paid to any thoughts that cultivators might be able to buy the land from its owners. It is to be expected that after the transitional phase most of them will face a situation where they no longer have access to land. Given the lack of non-agricultural wage-labour opportunities they will also have a rather weak position in negotiations with the landowners over new contracts.

It remains to be added that the members of parliament representing oppositional parties have warned of a new *fitna*, a war breaking out in the Egyptian countryside, besides that already existing between Muslim fundamentalists and the state. This time it would rage between the evicted peasants and the landowners. Fauziyya's husband (household 20) expressed a similar opinion when commenting on the parliamentary discussions before the passing of the law:

> No, I can't imagine that this law will be passed. The land is the bread of the people, they depend on it for their living. If they are robbed of it they will all go to the landowner's home and attack him. They cannot survive without the land.

With the passing of the new law, conditions have once again been rendered more favourable to the large landowners in the village and the old landowning aristocracy. These latter had re-emerged as a political force in the middle of the 1980s by allying themselves to the ruling party. In this sense Ansari, in an evaluation of the parliamentary elections of 1984, had already observed their recovery:

> There is no doubt that the overwhelming victory of the NDP was due to the support it received from the rural areas. Behind this victory was the NDP's mobilization of traditional elites . . . . [T]he NDP won the support of the majority of the rural notables because, in their eyes, it was the government party. (Ansari 1987; 245)

Ansari also mentions that some of these traditional elites were the direct descendents of pre-revolutionary parliamentary members going back to the last century (Ansari 1987: 245). Their rising influence and power in the ruling party have been very obviously demonstrated by the new law.

GLOBAL FLOWS AND THE CONSTRUCTION OF REALITIES

It would be a misconception to understand market integration as an incorporation of internally undifferentiated, static societal entities with clear-cut boundaries – a process by which national, rural and urban, traditional and modern entities, cultural, economic and political realms fit together. For one thing, village society is no amorphous, homogeneous mass, but consists of families, households, household members, neighbourhoods, landowners and the landless, entangled by conflicts and alliances, made dynamic by collective and individual strategies. It would likewise be wrong to understand market integration as the circulation of flows between such societal entities, as one would, for example, look at a network of railways or highways linking towns, cities, and states. I have been dealing in this book with the lives of the people 'inside a third world village', their economic activities and their ways of making sense of the world they live in. But this examination did not begin when I entered a certain territory, enclosed within clear-cut boundaries, where everything is different and oppositional from what lies beyond its boundaries, where the 'first', 'second', or 'fourth' worlds begin, and which are interconnected by means of mass communication, which I reached by taxi, by airplane, by taking the metro, then finally going by bus.

Market integration and the intersection of flows

Perhaps a somewhat more adequate conception of market integration would be one of a multitude of flows paralleling each other, intersecting, running in opposite directions, coming to a standstill, making fresh starts. These flows may be independent or dependent on each other, strengthening or weakening, or in conflict with each other. Equally important, flows are by no means purely economic, perhaps even not predominantly economic – that is, consisting of commodities. But they are at the same time cultural, transporting meaning, images, ideas, values, laws, narratives; they are financial, carrying along capital, remittances, debt, credits; they are technological and scientific; they consist of human beings.

The territory termed 'inside the Third World village' – as any other territory – may then be the very result of these flows, as well as the product of the social actors involved in 'indigenizing' (Appadurai 1990) flows: by taking them up, moulding and shaping them into spaces or territories, and into maps, at a certain historical juncture and a distinct geographical location. In this sense 'inside the third world village' is the outcome of the intersection of a multitude of flows, and the people's making sense and use of them – or ignoring them, complying with them, at times revolting against them. In this sense the absentee landlord, the state apparatus, the international development expert, the researcher, the village-based landowner, the remigrant, the priest, the daughters and mothers-in-law, all are involved in the creation and imagination of this distinct space.

For analytical reasons one may distinguish the local manifestations of structural flows from those which are the outcomes of the social actors' interacting and making use of such flows. In practice, however, distinctions between them are difficult to carve out.

Without the social actors' constructing reality – be this material or consisting of images and values – global market integration and its cultural, economic and political ramifications remain unthinkable. A large part of this book has therefore dealt with the fact that the village is in turn also the production site of flows: of agricultural produce, of labour power, of images.

This creation is obvious with regard to the more material and 'objective' consequences of flows: where government administrations have fixed a village's boundaries and *zimam* (the agricultural land belonging to a village); where it prescribes what has to be grown and sends employees to overlook the villagers' compliance with these regulations; where the government provides certain infrastructure facilities or not; where a remigrant from Saudi Arabia buys a used truck from the western part of Germany to bring limestone from Upper Egypt to the village (and for exporting rice straw, a village produce), a building material which changes the village's structure; where Italian noodle machines revolutionize the production of one of the major food staples; where western 'development technology' may lead to an altered form of water consumption; where a member of the government, an entrepreneur with family ties in the village, or the migrants have initiated substantial cash flows into the village.

Cultural flows and the construction of reality

The construction of territories, be these the Third World village, an urban central business district or an urban squatter settlement, is also dependent on different sorts of cultural flows. Evidently these are mostly transmitted by the media, especially TV programmes and its news and advertising, but also by the printed word and by films. It is in the cultural 'world cities' (Friedmann 1986) of Bombay, Cairo or Hollywood where human beings are engaged in the production of the material and immaterial content of such flows. An interesting case in point relates to the media reporting about the people who turned into refugees during the Gulf crisis. They were generally represented as impoverished, destitute and of rural origin.[7] Yet the production of cultural flows is not restricted to the media and film industry, as any kind of flow contains messages of meaning.

In the village community, western women like myself become transmitters of various images as I have shown in Chapter 3 of this book. When villagers described their pictures of the 'developed' world, the world of cities and centres of capital, I could often observe how both sorts of 'different flows' merge. They imagined me in the 'developed' world of my home society, an imagined world consisting of the products of TV advertising, of the cosmetic articles they discovered in my bag, of what I had told them about family planning, family ties and working women in my country. But flows often run in the opposite direction: people 'inside the first world city' wear underwear made of Egyptian cotton, visit Nofretete in Berlin, attend an Egyptian food festival and its supporting programme and go to a presentation of Egyptian folklore performed at a Mövenpick hotel; the less well-to-do may enjoy the belly dances performed at weekends at an Egyptian low budget restaurant. Thus the parallelism of different kinds of flows – of economic, human, cultural flows and their convertibility – becomes apparent as the small restaurant owner of the last example turns out to be an Egyptian migrant, and the customer of the Mövenpick a foreign businessman who feels at home particularly in Mövenpick hotels, which provide the same standards worldwide.

Apparently, their exists an intimate relationship between the content of cultural flows and the construction of the 'other' and its territory. Certainly, this book is only one other example in this respect. With all the 'facts' carefully composed throughout the chapters, the reader has pictured what seems to be the reality lying 'inside the Third World village', a construction constituted from all sorts of flows. In this way, 'inside the Third World village' has become 'real'.

Merging fiction and reality

Accelerating speed and a growing variety of forms are characteristic features of both market integration on a global scale and its constituent flows; this is obviously the result of the development of mass media and mass communication. In the last century it needed the Paris World Exhibitions to convey an image of Egyptian 'traditional' life to 'common people'. Mitchell (1988) has described in his *Colonizing Egypt* how material flows (of donkeys, donkey drivers, perfumes) were the building material of what was to become the representation of Egypt. In his concern with the construction of representations Mitchell has shown how boundaries between the fictional world of Egypt's urban life represented at the Paris World Exhibition and the reality lying beyond its premises were converging into each other. Visitors strolling around were getting confused as to whether they were still inside the exhibition or had already entered the residential, commercial and production areas lying beyond. Representation and reality became increasingly blurred. It seems that at the end of the twentieth century the infrastructure of mass tourism has tremendously accelerated and multiplied this mechanism. Moreover, representations and pictures of Egyptian traditional life are often produced in the Egyptian countryside itself. Here is a contemporary example of this sort of convergence. A local TV feature shows German tourists dressed in leisure clothes being welcomed with 'traditional hospitality' to a holiday resort by the natives. While there they ride indigenous donkeys (just as Mitchell has related about the World Exhibition), and enter a genuine peasant home or genuine bedouin tent. An Arab translator tells the local TV watchers of the tourists' enthusiastic narratives of having found the 'real life' of the Egyptians. Predetermined representations, coloured by experience from a distance, may have this effect when Egyptian peasant TV watchers and western tourists (or western female researchers, for that matter) accidentally meet in the streets of central Cairo or outside a construction site in Sharm al-Sheikh, South Sinai. That is, representations become the representations of representations; constructions the constructions of constructions.

Stable elements: the discourse on modernity and tradition

At any geographical–historical juncture of intersecting global flows, 'tradition' and 'modernity' seem to be present as vital constituents, and are perhaps even among its rare stable elements. This dualism can be tracked down in the village professional working in Cairo, and the prosperous, long-term resident migrant, both of them building their summer resorts in the village, the 'natural' environment; it is present in the village-based landowners' and the small peasant remigrants' construction of village tradition, its networks of reciprocity and its moral universe; and in that of mass tourists searching for authenticity; it is present in the physical appearance of the village; it is also there in the 'traditional' forms of production within the overall 'modern' (i.e., capitalist) mode of production. Moreover, it is also an easily noticeable aspect of scientific production, be this within the theoretical paradigms of modernity, neo-modernity or dependence, or in the nostalgic view of western critical social scientists (Stauth and Turner 1988).

This dualism is definitely also a constituent element of village discourse. Villagers use a certain style of dress on exceptional occasions and on holidays and thereby show that they know what a modern life-style is and how to behave according to it. Other villagers voice their disdain about the modern behaviour of a certain group of remigrants and their show-off behaviour. Different concepts of modernity and tradition are part of the discourse between the villagers and the western researcher as I have shown already in the first pages of this book.

The third world village, more often than not, is represented as the epitome of tradition. But the villagers of the 'traditional village' are not just staffage, the building material for outsiders' imaginations – they give expression to their own interpretations. They engage in the creation of images and meaning in their own right: modernity becomes development, cleanliness, frozen food, going to the hair-dressers, eating with knife and fork, the rationalization of bread production, the possession of a car. Yet simultaneously, moral decay, exaggerated individualism, loss of affection and security, defiance with religious beliefs and practices are seen as the embodiments of modern life. Tradition, on the other hand, signifies dirt, bad smells, poverty, illiteracy and illness, but also the place where one deeply feels at home, where *ʿishra* is still a characteristic feature of village society.

Flows coming to an end and re-emerging

Except for the endlessly recurring theme of 'modern' and 'traditional' life-styles, forms of production and territories, not much seems to be stable at the intersecting points of global flows. There may indeed no longer be a junction where only yesterday there had been one. And tomorrow new ones appear seemingly out of nothing. Mass tourist flows during the Gulf crisis came to a standstill, only slowly regaining momentum after the war. These were then far weaker than had been projected[8] by experts and hoped for by those engaged in the tourist business. At the beginning of the last decade of this century there is no longer any flow of Egyptian migrants to Iraq and no flows of remittances to the village, flows of *tarahil* labourers inside the national boundaries of Egypt are on the increase.

Coming to an end, flows are the building stones of what Appadurai (1990) has termed the 'landscapes' delineating the 'global cultural economy'. He has rightly made a point of the fluidity and historical situatedness of these landscapes when he clarified that he uses

> the (common) suffix scape to indicate first of all that these are not objectively given relations which look the same from every angle of vision, but rather that they are deeply perspectival constructs, inflected very much by the historical, linguistic and political situatedness of different sorts of actors: nation-states, multinationals, diasporic communities, as well as sub-national groupings and movements (whether religious, political or economic), and even intimate face-to-face groups, such as villages, neighborhoods and families.
>
> (Appadurai 1990: 296)

What Appadurai classified as salient features of the 'ethnoscapes' (as one of the five 'scapes' he conceptualizes as constituent of the global cultural economy) may be indeed inherent characteristics of all other kinds of flows where:

> the realities of having to move, or the fantasies of wanting to move . . . now function on larger scales. And as international capital shifts its needs, as production and technology generate different needs, as nation-states shift their policies on refugee populations, these moving groups can never afford to let their imaginations rest too long, even if they wished to.
>
> (Appadurai 1990: 297)

Egyptian migrants from small peasant households over the last two centuries, for that matter, could never have remained immovable and inflexible even had they wished to. They found themselves subjected to constantly shifting conditions of macro structures. Their work-force time and again became evaluated on working sites at different distances from their household, employed by different employers, sometimes of different nationality, performing different kinds of jobs, but still integrated in households holding to the ideal of shared production and consumption. I am afraid that to them the question as to whether over the past two centuries they constituted part of a globally articulated flow of migrants, of images and values of commodities, or remittances, is too academic and out of touch with reality to make much sense. For them, complexity is reduced to the desperate struggle to secure subsistence, of preserving a 'traditional' form of production and its moral defence – not because of some mystical essentialism and unreadiness for change, but because this is what is left to them as the most viable chance to secure the very reproduction of life. Being occupied with the construction of a confusing network of flows, this is a viewpoint we should not lose sight of.

NOTES

1 INTRODUCTION

1 See especially J.S. Birks *et al.*, 'Who is Migrating Where? An Overview of International Labour Migration in the Arab World', in Richards, A. and Martin, Ph.L., *Migration, Mechanization, and Agricultural Labor Markets in Egypt* (Boulder, Colo.: Westview Press, 1983); J.S. Birks and C.A. Sinclair, *International Migration and Development in the Arab Region* (Geneva: ILO, 1980); World Labour Report 1, *Employment, Incomes, Social Protection, New Information Technology* (Geneva: ILO, 1984).

2 The World Labour Report, op. cit., estimates 2.8 million in 1980. This is only one example of the discrepancy of statistical data.

3 See I.J. Seccombe and R.I. Lawless, 'State Intervention and the International Labour Market: A Review of Labour Emigration Policies in the Arab World', in Appleyard, R. (ed.), *The Impact of International Migration on Developing Countries*, pp. 73–4 (Paris: OECD, 1989), or D.A. Roy, 'Egyptian Emigrant Labor: Domestic Consequences', *Middle Eastern Studies* 27(4), 551–82, on the development of emigration policy in Egypt.

4 Unfortunately, Bendiab does not give the title of the work by Sylvain – another aspect of the unreliability of statistical material.

5 The typical loose, shirt-like garment of the rural male population in Egypt.

6 Here again is some evidence for the obstacles arising when trying to make use of statistical data. Methodological problems make it difficult to compare the findings referred to above as they are based on different units of analysis. Some comprise the economically active population, others count households with migrants at the date of surveying, while still others include those with return migrants. Our own experience gives hints to other problems: for example, as interviewing by using questionnaires always raises the respondents' doubts about the usage of the data (i.e. their fear that some government institution might get hold of it and use it to extort money), there certainly are households which avoid mentioning the existence of a migrant member.

7 This book presents the outcomes of my own research within a larger migration project comprising six rural communities, four of them situated in Lower Egypt, two others in the Middle Egyptian governorate of Minia. The project was a joint research between a team of the Social Research Center of the American University in Cairo and of the Sociology of Development Research Center of Bielefeld University in Germany. The study was funded by the German Research Foundation (Deutsche Forschungsgemeinschaft). Quantitative material collected during the field research and partly analysed in the present study, consists of different sets of data:

1 A standardized household census survey was conducted in all of the six villages, providing basic socioeconomic data. While totalling n = 8,620 households, the corresponding number for Kafr al-ʿIshra is 287.

2 In the census 2,483 migrants were identified, of which 1,765 had returned by the time of interviewing. From the returnee migrants random samples for each village were selected, representing a total of n = 639 cases. In their households extensive semi-standardized interviews were conducted. In Kafr al-ʿIshra 100 households were questioned.

Interviews were conducted by secondary school and university students and graduates residing in the villages, under the guidance of the research teams.

2 EGYPT'S RURAL AREAS OVER THE PAST TWO CENTURIES

1 In mid-1989 government officials estimated that unemployment ranged between 20–22 per cent. Given an inhabited and cultivated area of 35,200 sq. km, and a population of 56 million (1991), population density is 1.590 capita per sq. km. As a comparison: Northrhine Westphalia, the most densely populated federal land of Germany, has a population density of 500.

2 The reader is instead referred to G. Amin and E. Awny, *International Migration of Egyptian Labour. A Review of the State of the Art* (Ottawa: International Development Research Centre, 1985).

3 See N. Choucri, 'The New Migration in the Middle East: A Problem for Whom?', *International Migration Review* 11(4) [1977], 421–43; and ibid., 'Asians in the Arab World: Labor Migration and Public Policy', *Middle Eastern Studies* 22(2) [1986], 252–73. As he wrote: 'To a large extent, the Arab Middle East is a closed system in that demographic characteristics have not been influenced by large-scale out-migration' (1977: 422). The term in itself is highly disputable as a closed system would refer to a state where neither in- nor out-migration was possible. Otherwise the term 'semi-permeable' would have been more appropriate. What is more, while including the Maghreb states in his definition of the 'Arab Middle East', Choucri does not mention the

large-scale out-migration to the northern shore of the Mediterranean and thus his notion of a 'closed system' loses any meaningfulness. About ten years later Choucri dealt with the massive in-migration of Asians to the Gulf region (Choucri 1986), which although already an important fact when he wrote his first article had not then found his attention. But even here, while now acknowledging the fact that already in the second half of the 1970s 'Asians were now almost as important as Arabs in the region' (1986: 254), this does not lead him to any major revision of the concept. While no longer speaking of a 'closed system' it appears that Asians are simply integrated into what is now termed the 'Arab world'. My point is that if one highlights the overwhelming and ever-rising importance of Asian labour power in that region one also somehow has to come theoretically to terms with it which inevitably must lead to the conceptualization of some broader system.

4 See J.H. Brink, 'The Effect of Emigration of Husbands on the Status of their Wives: An Egyptian Case', *International Journal of Middle East Studies* 23(2), 201–11; M. Hamman, 'Labour Migration and the Sexual Division of Labour', *Middle East Report*, no. 95, 5–11; F. Khafagy, 'Women and Labor Migration: One Village in Egypt', *Middle East Report*, no. 124, 14–21; H.A.S. Khattab and S.G. El Daeif, *Impact of Male Labor Migration on the Structure of the Family and the Roles of Women* (Cairo: The Population Council of West Asia and North Africa, 1982); and E. Taylor, 'Egyptian Migration and Peasant Wives', *MERIP Reports*, no. 124, 3–10.

5 I am indebted to Butrus Abu Manneh, Haifa, for this information.

6 For a description of the organization of an *'izba* at the end of the nineteenth century see J.F. Nahas, *Situation économique et sociale du fellah égyptien* (Paris: A. Rousseau, 1901); and in the 1940s S. Saffa, *Exploitation économique et agricole d'une domaine rural égyptien* (Le Caire: Thèse, 1949). See on the *'izba* as a system of production G. Stauth, 'Zur Auflösung und Peripherisierung integrierter Kleinbauernwirtschaft – Das Beispiel der Fellachen im Nildelta', in *Arbeitsgruppe Bielefelder Entwicklungssoziologen* (eds), *Subsistenzproduktion und Akkumulation* (Saarbrücken/Fort Lauderdale: Breitenbach, 1981); ibid., *Die Fellachen im Nildelta* (Wiesbaden: Steiner, 1983), ibid., 'Capitalist Farming and Small Peasant Households in Egypt', in Glavanis, K. and Glavanis P., *The Rural Middle East. Peasant Lives and Modes of Production* (London and New Jersey: Zed Press, 1990). The original meaning of *'izba* is that of an 'encampment of temporary straw huts put up for labourers working at some distance from their village'. (See R. Owen, The Middle East In The World Economy 1800–1914, London: Methuen.) referring to J. Lozach and G. Hug, *L'habitat rural en Egypte* (Cairo: Société de géographie d'Egypte, 1930.)

7 But it was by no means a uniquely Egyptian phenomenon, as Richards has pointed out, comparing it to the East Elbian *Gutswirtschaft* and the *inquilinaje* system of Central Chile. (See A. Richards, 'The Political Economy of Gutswirtschaft: A Comparative Analysis of East Elbian Germany, Egypt and Chile', *Comparative Studies in Society and History*, vol. 21, 1979 483–518; and also Stauth, op. cit., 1990.) Von

Werlhof describes the *conuco* system of Venezuela, a similar institution. (See C. von Werlhof, *Wenn die Bauern wiederkommen* (Bremen: Ed. Con., 1985).)

8 See D.M. Wallace, *Egypt and the Egyptian Question* (London: Macmillan & Co., 1883), pp. 180–7, for an impressive account of the increasing indebtedness and loss of land of a formerly self-sufficient peasant household due to heavy taxation and dependency on money lenders.

9 Cotton was produced especially in Lower Egypt, whereas in Upper Egypt sugar-cane as a cash crop dominated.

10 It has to be noticed, though, that the images of a Pharaonic identity are still predominant in Egypt's tourist industry and that for some reason or other the image of the bedouin also seems to be more marketable than that of the peasant. Yet, in the endeavour to create new markets, the peasant imagery is also appearing as another focal point in the search for the original and authentic. Thus, new types of postcards representing the Egyptian countryside and its inhabitants are gradually emerging and are appealing to the aesthetic preferences of the educated tourist (*Bildungstourist*). And also visits to a real village are increasingly arranged by the animators working in the tourist villages.

11 On rural exodus see B. Hansen and S. Radwan, *Employment Opportunities and Equity in a Changing Economy: Egypt in the 1980s, a Labour Market Approach* (Geneva: ILO, 1982), p. 93; J. Abu Lughod, 'Migrant Adjustment to City Life: The Egyptian Villages in Cairo', *Ekistics*, vol. 25, 192–202; G. Sabagh, 'Migration and Social Mobility in Egypt', in Kerr. M.H. and Yassin, S. (eds), *Rich and Poor States in the Middle East. Egypt and the New Arab Order* (Cairo: American University in Cairo Press, 1982), p. 76; and Saffa, op. cit., pp. 417–18.

12 Qat is a smooth-stemmed shrub (*Catta edulis* or *methyscophyllum glaucum*), the leaves and young shoots of which contain an alkaloid (katin) which produces a euphoric and stimulating, but finally depressing, effect. It is very widely chewed in Yemen. This is a social ceremony which takes place in the afternoons; it has won such enormous popularity that economic life virtually comes to a standstill at that time of day. For more information, see the article on qat in E. van Danzel, B. Lewis and Ch. Pellat, *The Encyclopaedia of Islam* (Leiden: E.J. Brill, 1978), vol. IV, p. 741.

13 The respective figures for women are 15.9 per cent for Egyptian women and 21.5 per cent for Kuwaiti women. Indian, Pakistani and Bangladeshi women formed 20.3 per cent of the female work-force according to the survey of 1983.

14 There are certainly other reasons for the preference of Egyptians over other Arab nationals; among others, political reasons must be cited.

3 TRACKING DOWN GLOBAL FLOWS IN THE THIRD WORLD VILLAGE

1 In the metropolitan context one of the most apparent examples recently has become the mobile telephone, whose use value is not so much that

of being always within reach but that of showing off a yuppie status and life-style.

2 As for example when welcoming a rare or high status visitor at joyous festivities, after a promotion or on weddings. In the village people used to say when they wish good luck to somebody having to take an exam: 'So God will we shall drink Coca-Cola after your school exam!'

3 See the quotations by Paye in the preceding chapter (p. 19).

4 For the sake of anonymity all names have been changed.

5 Pope Cyril (Kirollos) IV, pope and patriarch of Alexandria and All-Africa from 1959–1971.

6 The nightdress is the typical summer indoors dress of urban Egyptian women. Visiting friends and relatives are often offered to change their outdoor dresses for such clothing to feel more relaxed. While in the hot season one definitely feels more comfortable in the nightdress, at the same time it is an expression of familiarity and hospitality. In the village the traditional female *gallabeyya* is offered rather than a nightgown – I suppose that many of the village women simply do not dispose of such a ready-made nightdress.

7 Saint George of Capadocia, born in AD 280. His grave is in Mit Damsis, Egypt. Many Egyptian churches are named after him.

8 My whole stay in the village was accompanied by this sort of discussion. Clearly, the issue of having children remained the most important topic in this discourse. However, I felt that while we came closer together over time, by developing feelings of friendship and affection, interests changed: these women, convinced of the correctness of their arguments and experiences, urged and tried desperately to convince me of the importance of becoming a mother because they wanted me to be happy.

9 I must add that I was really shocked about the more material effect of this kind of irresponsible product promotion. The people made use of these insecticides as demonstrated in the representations. Not only were enormous quantities used, but what frightened me most was how carelessly the poison was sprayed on nearly everything. Nowhere was there any warning about adverse health effects, the right usage, and that this after all was not a deodorant but a poison.

10 I want to go into further detail here about the discussion which took place during my first day in the village. This struggle over power I was involved in had other vital dimensions. One was the personal interest of Samia who saw in her task of introducing me to migrant families a very welcome chance to escape from unpleasant household scores or to visit those persons she wanted to have a chat with. Through me she managed to win herself an extraordinary freedom of movement as she could even leave the village and visit not only the district town but also the governorate's capital. This later developed into a real conflict between the two of us about whom we were to visit – the persons I wanted to meet or the families she proposed to me. I felt I could only resolve this conflict in my favour by bringing in the ʿOmda's authority.

Another issue I want to mention briefly here is the religious identity and practices of Im Ayman, introduced earlier in this chapter. These, as I gradually noticed, by far exceeded those of her Christian compatriots. While the community's priest definitely is the undisputed Christian authority, I got to know Im Ayman as a woman living on her own who is engaged in establishing for herself a position as a woman expert in popular religion. This self-stylization was obvious in the extensive decorating of her house, the courtyard, and even, on occasion, the alley in front of it, with pictures of Christian saints. Thus it was no surprise to me to find her being consulted by village women with ill children.

4 KAFR AL-'ISHRA: AN AGRARIAN VILLAGE IN THE 1980s

1 On the introduction and legislation of private landed property see G. Baer, *A History of Landownership in Modern Egypt, 1800–1950* (London: Oxford University Press, 1962); K. Cuno, 'The Origins of Private Ownership of Land in Egypt: A Reappraisal', *International Journal of Middle East Studies* 12(3), 245–75; and A. Richards, *Egypt's Agricultural Development, 1800–1980. Technical and Social Change* (Boulder, Colo.: Westview Press, 1982).

2 Another typical subject of national soap operas is the Upper Egyptian village and the 'Omda's household. This time the stage is set for the images of the village way of life and its traditional values.

3 N = 236. In the other five villages we did research in an average of 66 per cent of the buildings were already constructed of red brick.

4 It is forbidden due to the acute shortage of agricultural land.

5 In the other five villages this percentage has been far lower with 9.2 per cent.

6 I want to elaborate on this a little further. I very often have heard villagers complain about the necessity to enrol their children for private lessons. Parents and pupils reproach teachers for not explaining the curriculum during the regular classes but only in the private lessons in the afternoon; this means that all those who want to succeed in important exams are forced to pay considerable sums to attend private teaching. Villagers say that it is especially the young teachers wanting to marry who engage in this practice, that if they do not get the chance to migrate they try to cover marriage expenses by giving these private lessons. The aforementioned, illuminates the unreliability of statistical data. Learning that a certain percentage of children do not attend primary school does not really give us a true picture. I know quite a lot of people without any formal education who are able to write and calculate whereas there are many whose five years of education have not resulted in even the slightest skills of this kind.

7 According to the Constitution 'Islam is the religion of the State' and the Islamic code is a principal source of legislation. The state safeguards the freedom of worship.

8 'Week of awakening or resurrection'. During this week Christian villagers wait for the appearance of the saint or his spirit.

9 See, however, the changes in the relationship between cultivator and landowner introduced by a new law in June 1992 mentioned in the final chapter.

10 This indicates that two partners join in the 'cultivation of the half'. Though A. Richards and Ph.L. Martin state in *Migration, Mechanization, and Agricultural Labor Markets in Egypt* (Boulder, Colo./Cairo: Westview Press, 1983, p. 4), that in Egypt most rents are in cash and share-cropping is not common this is not the case in Kafr al-ʿIshra where *muzaraʿa bin-nuss* is the predominant form of rent.

11 Sometimes *muzaraʿa* contracts of this type are connected with the *bin-nuss* system. This can be the case when the landowner is not involved in agriculture and living in the city. They are most interested in the growing of crops which assure a monetary return. Peasant households, on the other hand, are more interested in the production of crops which enter the consumption of the household (e.g., *barsim*, clover). Then sometimes arrangements are concluded whereby the cultivator rents the landowner's half for the season and grows clover on it.

12 A *qantar* equals approximately 45 kg.

13 Though complaints about poor monetary returns especially in the case of cash-rents, are frequently to be heard. Meagre net returns have in fact induced members of the old absentee class in particular to sell their land. With regard to the advantages of the system I am referring to the village-based big and middle landowners.

14 This, however, no longer seems to be the case with cotton growing which peasants think to be detrimental to their subsistence in any case; thus any additional input of labour power and fertilizers, etc. is regarded to be wasteful.

5 KAFR AL-ʿISHRA'S PEASANT HOUSEHOLDS

1 That is, a person who had formerly migrated and already returned or a member currently abroad. This is astonishing as the percentage of the other five villages we studied and where out-migration began much earlier was only 26.8 per cent.

2 However, the argument of a deficit of labour power filled by Egyptian migrants remains questionable to me as I have already explained in Chapter 2 with regard to Indians working in Yemen. In Jordan, frequent complaints about the Palestinian indigenous labour force point to the fact that Egyptian migrant labour, which is far cheaper than the local labour force, serves to bring down wages and therefore may be a serious competitor over scarce jobs. This finds an expression in latent racism between Egyptians and Palestinians. It is not surprising that in one of the migrant accounts about experiences abroad (see p. 110) the same motive occurs. It would be interesting to study the thesis that by accepting Egyptian migrants, small and capital-weak enterprises secure their survival in a situation of heavy competition by employing an exceptionally cheap labour force (see A. Portes and J. Walton, *Labor,*

Class, and the International System (New York: Academic Press, 1981), p. 56).

3 For a similar account see F. Khafagy, 'Socio-Economic Impact of Emigration from a Giza Village', in Richards, A. and Martin, Ph.L., *Migration, Mechanization, and Agricultural Labor Markets in Egypt* (Boulder, Colo./Cairo: Westview Press, 1983), pp. 142–154.

4 Short-term migration is also a feature of the other villages we studied, here migrants stayed an average of 2.7 years (mean for the five other villages).

5 While this was generally also true for the five other villages (78.5 per cent of the migrants from these villages do unskilled/skilled labour), their percentages, nevertheless, have been somewhat more favourable.

6 The average for the five villages, with 13.2 per cent, is again more favourable.

7 See Map 5.1, p. 121.

8 The questionnaire included questions asking for earnings abroad although it had been anticipated that answers would hardly be reliable, given the suspicions of the villagers to disclose their financial situation to outside forces whose relationship to the government remains doubtful. The given results confirmed this and therefore were not evaluated. A more reliable procedure was to concentrate on remittances spending. Here Kafr al-ʿIshra was clearly the village where the least remittances seem to have been at the disposal of the return migrants.

9 The average for all the six villages under study is also 0.7 per cent.

10 The number in parentheses indicated the number of the household according to Map 5.1.

11 1, 5, 9, 11, 16.

12 The households which lost children are 1, 3, 4, 5, 7, 8, 9, 15, 16, 18, 20; those who reported that none of their children had died are 12, 17, 19. The woman of household 17 was only married a relatively short time and after being divorced never married again. I have no information about child death in households 6 and 11.

13 Households 3, 4, 5, 8, 9, 12, 16.

14 This sort of diploma is acquired by those who after primary (five school years) and preparatory school (three years) have continued their education in a technical, agricultural or administrative institute (three years).

15 1, 3, 4, 6, 8, 9, 11, 15, 16, 18–20.

16 Five households (4, 6, 8, 9, 11) have access to owned and rented land, four (3, 15, 18, 16) have only owned land, and one (20) has only rented land.

17 1, 5, 7, 12, 17, 19.

18 1, 7, 19.

19 A *kila* equals 16 kg (i.e., 128 kg of wheat).

20 5, 15, 17, 12.

21 Entitling her to half the produce without obliging her to forward any inputs.

22 1, 3, 4, 5, 7.

23 According to our survey, in Kafr al-ʿIshra only 7.7 per cent of all persons were under 20 at the time of their first migration, whereas 85.4 per cent were between 20 and 39 years old.

24 See Chapter 4, the social geography of the village.

25 Labour rent is defined as resulting from the free transfer of the labour force of the household into the capitalist sector. Surplus value comes from the exploitation of wage labour within a capitalistic enterprise.

26 For a review D. Wong, *Peasants in the Making, Malaysia's Green Revolution* (Singapore: ISEAS, 1987), ch. 2; and K. and P. Glavanis, 'The Sociology of Agrarian Relations in the Middle East: The Persistence of Household Production', *Current Sociology* 31(2), 4–43.

27 Bennholdt-Thomsen even presumes that strategies for the revival and strengthening of subsistence production through international agencies like the World Bank, UNO or ILO have also to be understood in this sense. (See V. Bennholdt-Thomsen, 'Investition in die Armen. Zur Entwicklungspolitik der Weltbank', *Lateinamerika, Analysen und Berichte*, vol. 4, 74–96.

28 Wallerstein defines the labour force's variability in time and space, as well as its low costs, as the primary concern of the accumulators. (See I. Wallerstein, 'Household Structures and Labor-Force Formation in the Capitalist World-Economy', in J. Smith, I. Wallerstein and H.D. Evers (eds), *Households and the World-Economy* (Beverly Hills, 1984, my translation), p. 18.)

29 Here are two characteristic quotations from Meillassoux:

> We will not have to examine the destruction of one mode of production by another, but the contradictory organization of economic relationships between two sectors, the capitalist and the domestic, whereby one sustains the other in order to deprive it of its substance and thereby destroys it (Meillassoux 1976: 116).

In another place he says that

> The 'domestic mode of production' consisting of homologous communities sustaining organic relationships with equal communities does no longer exist .... The 'domestic community' – squeezed, extorted, dismembered, registered, taxed, recruited – is swaying, but is still holding ground, because the domestic relations of production have not totally vanished. Millions of production cells are still organized in this form, which are integrated into the capitalist society to a differing degree and are voided of their substance and energy by the bruising weight of imperialism (Meillassoux 1976: 106; my translation).

6 MIGRATION AS A STRATEGY OF THE SMALL PEASANT HOUSEHOLD

1 For a Sudanese village, see also V. Bernal, *Household Agricultural Production and Off-Farm Work in a Blue Nile Village, Sudan* (Evanston, Ill.: Northwestern University, 1985); and ibid., *Cultivating Workers. Peasants and Capitalism in a Sudanese Village* (New York: Columbia University Press, 1991). For a similar opinion see K. and P. Glavanis,

'The Sociology of Agrarian Relations in the Middle East: The Persistence of Household Production, *Current Sociology* 32(2), pp. 19–20, 39; and Stauth (1981).

2 Only households with a landowning larger than 3 *faddan* are considered to have an income from farming alone which exceeds poverty-line income. Radwan and Lee point to the high variation in the amount of cash 1 *faddan* can generate and state that the net value of farm output of households in the less than 1 *faddan* category may be between 8 per cent and 46 per cent of the total income. See S. Radwan and E. Lee, *Agrarian Change in Egypt. An Anatomy of Rural Poverty* (London: 1986), p. 62; also A. Richards, *Egypt's Agricultural Development, 1800–1980. Technical and Social Change* (Boulder, Colo.: Westview Press, 1982). p. 2.

3 Mechanization of Kafr al-ʿIshra also seems to have negative impacts on the small and poor households. An example is the introduction of diesel water pumps which led to the neglecting of the *tambusha*, the animal-power-led waterwheel which had been used until then (and in turn had replaced the Archimedes' screw). As these waterwheels are practically all out of order in the village today, households which do not have their own water pump must rent one. At the time of field research, the price of irrigation of 1 *faddan* was £E12. As rice must be watered every second day and maize every eighth to tenth day this poses considerable additional cash demands on the household's budget. Threshing is another case. This work is today totally mechanized in the village and nobody exclusively uses animal and manpower any longer.

4 It is this dimension which also allows us to assess the relative success of the migrant who went to Austria, as he was able to buy a whole *faddan* of land.

5 It has to be mentioned here that buying a buffalo is often not a real enhancement of the household's economic standing, as the travel itself is often financed by the selling of a cow, as we will see in the next chapter. Buying a buffalo in these cases is only to regain the pre-migration status.

6 The mean for the other five villages is 31.35 per cent.

7 Between the decade of the two population censuses of 1976 and 1986 the average annual growth rate was estimated at slightly below 3 per cent. In 1976, in the governorate of Daqahliyya to which Kafr al-ʿIshra belongs, 788.6 inhabitants per square km were counted, and their number had risen by 1986 to 1,008.5 (see Statistisches Bundesamt, *Länderbericht Ägypten* (Wiesbaden, 1988).

8 Which however does not impede profit orientation of households which own land in the vicinity of the inhabited area. Though forbidden, it is no secret that many sell their land as construction sites and not as agricultural land, and thereby made considerable profits. The price for a *qirat* of construction land in 1987/8 in Kafr al-ʿIshra was £E6,000.

9 Even if the space is available, then financing poses problems. This in fact is also the case for the better-off peasant households. I remember the widowed woman of the late village headman of Kafr al-ʿIshra (a position which is linked to the possession of at least 10 *faddan*

agricultural soil), with a son at marriageable age, say: 'but *who* can marry his son and build additional rooms or an additional floor at the same time? I can only finish the building and then perhaps in the next year, if God wills, I can think about looking for a bride for my son!'

10 I have no precise figures for 1988, but in 1992, when I returned to the village, pupils had to pay to the private teacher £E5 per subject per month.

11 And who would have been uneducated. The bed and the wardrobe or a trunk are the traditional items of the *gihaz*, the bride's trousseau.

12 Another important reason for the purchase of gold is the repayment of loans, as the wife's gold is often sold to finance the travel abroad, as we will see in the next chapter.

13 Or in Egyptian dialect, a *hagg*. See, as an example, household 18 (Map 5.1).

14 Here is an example. A young woman had sold all her gold in order to help her husband to open a chicken farm which was meant to be the economic basis of their household. After the business became established, he did not buy her gold back or buy household appliances but bought a motorbike for himself, which, as he explained to his wife, was an absolute necessity for his project. This was in 1988. When I came back in 1992 I learnt that he had not been successful in his chicken enterprise and had eventually given it up with considerable loss, and that for over a year he had been engaged in rabbit farming. In order to get this project off the ground his wife even sold her wedding ring. But though in business for over a year they had not earned anything out of it. His wife was very reluctant and said that he would never be successful in it. I did not notice anything they had bought for the household during these years. Yet, the husband still had his motorbike.

15 For a critique, see N.J. Brown, *Peasant Politics in Modern Egypt. The Struggle Against the State* (New Haven, Conn. and London: Yale University Press, 1990), p. 59; and my example of the urban bus driver taking for granted that those who can't drive a car must be peasants (Ch. 3).

16 Ibrahim's actions did, however, not remain without comment in the two streets. As he is from a rather poor household, his marriage did not conform with the maxim not to weaken or endanger the household's potential of reproduction. Here is the comment of a woman in the two streets about Ibrahim's behaviour: 'I don't know how people can be so stupid. Now he has married a woman who is "better" than we are. She has a diploma and comes from the city. So she cannot even work in the fields. They have sold the *gamusa* and have nothing to live on. His parents only have another mouth to feed!'

17 Here is an interesting example of the link between wearing jewellery and high education. When I came back to the village in 1992, having finally acquired my Ph.D., the mother of Majid and Muhammad (5) taught me a lesson about the convertibility of economic and cultural capital. (For an in-depth discussion on types of capital and their significance, see P. Bourdieu, 'Ökonomisches Kapital, kulturelles Kapital, soziales Kapital', in Kreckel, R., *Soziale Ungleichheiten, Sonderband 2*

der Sozialen Welt (Göttingen: Schwarz, 1983.) As a matter of fact, I wasn't very concerned about my new status. However, being aware of the fact that virtually nothing about me could be kept secret, and knowing about the extraordinary importance of having such a title, I got the impression that villagers simply did not believe that I had become a 'doctor', as they only mentioned it rarely. Then, Im Majid made me understand: 'Look, in our society, any woman who becomes a *duktura* will very soon wear lots of golden bangles. And when she has gone to work in Saudi Arabia only for a year, she will buy lots of gold so that everybody can see that she is a *duktura*. And, of course, she also must return with a car . . . . This is definitely a must in our society – I know, not in yours. But here, you have to do this.'

18 Ibrahim's father, it is interesting to note, is of black origin, and Ibrahim also is rather dark.

19 3, 4, 7, 8, 10, 12, 18–20; eight, to be precise, as 10 is only partially finished.

20 And, to a still larger degree, for the daughters-in-law.

21 See, on the roles and status of village women: M. Abaza, 'The Changing Image of Women in Rural Egypt', *Cairo Papers in Social Science* 10(3); H. Ayrout, *The Egyptian Peasant* (Boston, 1963); U.B. Engelbrektsson, *The Force of Tradition: Turkish Migrants at Home and Abroad* (Goeteborg: Acta Universitatis Gothoburgensis, 1978), p. 127; D. Kandiyoti, 'Sex Roles and Social Change: A Comparative Appraisal of Turkey's Women', *Sign* 3(1), 57–73; V. Maher, *Women and Property in Morocco: Their Changing Relation to the Process of Social Stratification in the Middle Atlas* (London: Cambridge University Press, 1974); P. Weyland, 'Remigration and Geschlechterverhältnis auf dem ägyptischen Dorf. Forschungsperspektiven und hypothesen', Working Paper no. 95, Bielefeld, 1987; and S. Zimmermann, *The Women of Kafr al-Bahr. A Research into the Working Conditions of Women in an Egyptian Village* (Cairo and Leiden: Research Centre, Women and Development, State University of Leiden, Institute for Social Studies, 1982).

22 A woman is called *ghadbana* when she leaves her husband's household after a quarrel and goes back to her father's house. Women may stay away for some months without returning to their husbands, who eventually have to come and fetch them. A solution is often reached when the husband, intent on negotiating his wife's return by giving her a gift, visits his father-in-law. It is only after such an agreement is reached that the husband sees his wife again and can take her back to the matrimonial household.

23 A 'clever' peasant woman is one who possesses the virtuosity of 'making something out of nothing', of slowly increasing the household's resources. Sausan (household 17) serves as an other example in this respect. See also H. Ammar, *Growing up in an Egyptian Village* (London: Routledge, 1954), p. 22.

24 That the relationship between mother and son, especially in the first years of marriage, is often closer than the one between the spouses, is no secret (Ayrout, op. cit., p. 128; Kandiyoti, op. cit.) One of the married sons of an extended household I frequently visited put the issue

in a nutshell; when in the course of a quarrel with his wife he exclaimed: 'The woman who brought me up – how could I ever annoy her? But you – why shouldn't I?' The problems between them which had their origin in the young woman's complaint about overwork in the extended household soon reached new dimensions when it began to involve the senior couple. In the course of events, the mother-in-law went as far as requiring her son to leave the matrimonial bed and to stay with his still-unmarried brothers. She commented on her demand by saying: 'After all he is *my* son and you are only his wife!'

25 Actually, the battlefield is not only to be found inside the village. As different villages have migration stories of varying success, the battleground also has an inter-village and a rural–urban dimension. See also, G. Stauth and P. Weyland, 'Transformationen ländlicher Gesellschaft und sozialer Konflikt: Das Beispiel Ägypten', in Stauth, G., Albrecht, R., Reichert, Ch. and Weyland P., 'Fellachen zwischen Überlebenssicherung und Modernität als Lebensstil. Strukturelle Unterentwicklung, Migration, Geschlechterverhältnis und Islamisierung: Die Konflikte der ländlichen Gesellschaft Ägyptens, 1986–1988', Unpublished ms., Bielefeld, 1990.

7 ORGANIZING THE HOUSEHOLD'S SURVIVAL DURING THE MIGRANT'S ABSENCE

1 See H. Friedmann, 'Household Production and the National Economy: Concepts for the Analysis of Agrarian Formations', *Journal of Peasant Studies* 7(3), 158–84; and for a critique, D. Wong, *Peasants in the Making. Malaysia's Green Revolution* (Singapore: ISEAS, 1987).

2 The two remaining persons indicated skilled and clerical work as their primary jobs after return.

3 The database here is that of the remigrants of the household survey; this result is also corroborated by the remigrant sample. Here 49 out of 51 (96 per cent) indicated direct involvement in agricultural production *before* migration, whereas 51 out of 55 (93 per cent) did so after return.

4 Data from the migrant sample.

5 Mannheim defined cooperation in the following way:

> The most spontaneous co-operation between groups is the mutual aid between neighbours, which is a spontaneous combination of efforts without submission to authority. The characteristic of this spontaneous readiness to give mutual aid, is that it works better in hard times than in easy times . . .
>
> Generally speaking the lower social classes are more in favour of mutual aid than the middle or the upper classes. The explanation for this is that in hard times most people realize how essential are the principles of co-operation, mutual aid, the principles of *do ut des*.
>
> (Mannheim 1957: 92)

See on cooperation with respect to Egypt: M. Abaza, 'La paysanne égyptienne et le "féminisme traditionel"', *Peuples Mèditerranéens*

41–42, 135–51; H. Ammar, *Growing up in an Egyptian Village* (London: Routledge, 1954); and G. Stauth, *Die Fellachen in Nildelta* (Wiesbaden: Steiner, 1983).

6 *Shirk* ('partnership') is an institution whereby two or more households share in the costs and labour needed to raise or hold a cow. Usually poor households lacking the necessary amount of cash to buy a calf or a cow sometimes take recourse to *shirk*. There are several different forms of *shirk* practised in the village. In the predominant one, a rich household or a rich person buys the cow and the poorer has to maintain it, being responsible for fodder, cleaning of the stable and for milking it. In turn the poor household can dispose of the milk and gets half of the price if the cow has a calf which is sold. However, if the *gamusa* is sold the money goes entirely to that household which had initially bought it.

7 See, for example, J.H. Brink, 'The Effect of Emigration of Husbands on the Status of their Wives: An Egyptian Case', *International Journal of Middle East Studies* 23(2), 201–11; H.A.S. Khattab and S.G. El Daeif, *Impact of Male Labor Migration on the Structure of the Family and the Roles of Women* (Cairo: The Population Council of West Asia and North AFrica, 1982); and C. Myntti, 'Yemeni Workers Abroad: The Impact on Women', *Middle East Report*, June, pp. 11–16.

8 AFTER THE GULF CRISIS

1 According to E. Hooglund, in 'The Other Face of War', *Middle East Report*, vol. 21, no. 171(4): 'Kuwait, Saudi Arabia, Qatar and the UAE rewarded Egypt financially by providing outright cash grants, forgiving more than $7 billion in Egyptian military debt, and the Paris Club of Western creditor governments wrote off another $10 billion' (Hooglund 1991: 7).

2 See also the stories of remigrants about their time abroad in Chapter 5.

3 This is a quotation recorded during the field stay in 1987/88, therefore the Iranian-Iraqi war is alluded to.

4 This example struck me even further when I remembered the usage of the very same machine in certain circles of my own society. I do not know anything about the significance of noodle machines in Italy, yet in Western Germany they definitely indicated the growing desire to rediscover the domestic sphere of subsistence production, an endeavour which is linked not only to *more* work (opposite to the Egyptian case) but which is also immediately taken up by industry which, by putting these machines on the market has already turned the desire of the user to counteract the 'disembedding mechanisms' or the 'deskilling of day-to-day activity' to own profit. See A. Giddens, 'Structuration Theory: Past, Present and Future', in C.G.A. Bryant and D. Jary (eds), *Giddens' Theory of Structure. A Critical Appreciation* (London: Routledge, 1991), p. 207.

5 The *sabahhiyya* is a present of the bride's family and her friends delivered on the first morning of married life. It consists of everything the young housewife needs to start her own household: rice, wheat, fruit, vegetables,

sweets, soap, a pair of chickens or geese. While addressed to the bride, this is more a courtesy of her family to that of her husband's into which she is now integrated as the mother-in-law usually takes it for granted that everything is brought for her to dispose of as she wishes.

6 Law 96/1992 for 'the amendment of some paragraphs of law no 178/1952,' published in *al-jarida ar-rasmiyya* 26 (28 June 1992).

7 One example is a photo by Angel Franco (NYT Pictures; published in 1991 in *Middle East Report*, no. 171, p. 2) which is subtitled 'Egyptian refugee on Iraq/Iran border'. It shows a man, obviously an Egyptian peasant clothed in a worn-out, tattered *gallabeyya*, barefoot, somewhat scared, and with a sad, disappointed, tired expression on his face. This expression is, however, counterbalanced by a lightly clenched fist. The Egyptian peasant migrant, miserable as he appears, is used to performing hard physical work. See for a similar convergence of flows of people, images and meaning the cover picture of no. 168 (1991) of the same journal.

8 To create still greater confusion I want to mention the fact that there are also imagined future flows and standstills, creating self-fulfilling prophecies (Watzlawick 1985). Here the 'natural' flow from cause to effect is reversed. In the example mentioned the falsely projected future mass flows of tourists – a compensatory phenomenon for leisure time lost during the Gulf War – led to dramatic overcapacities on airline charter flights and in hotels.

BIBLIOGRAPHY

Abaza, M. (1987) 'The Changing Image of Women in Rural Egypt', *Cairo Papers in Social Science* 10(3).

—— (1988) 'La paysanne égyptienne et le "féminisme traditionel"', *Peuples Méditerranéens* 41–42: 135–51.

Abd al-Mu'ti, B. (1986) 'L'emigration et l'avenir de la question sociale en Égypt', *Revue Tiers Monde* (jui–sept).

Abdel-Fadhil, M. (1975) *Development, Income Distribution and Social Change in Rural Egypt 1952–1970*, Cambridge: CUP.

—— (1979) *Oil and Arab Unity* (in Arabic), Beirut: Center for Arab Unity Studies.

—— (1980) *The Political Economy of Nasserism: A Study in Employment and Income Distribution Politics in Urban Egypt, 1952–1972*, Cambridge and New York: CUP.

Abdel-Khalek, G. (1981) 'Looking Outside, or Turning Northwest: On the Meaning and External Dimensions of Egypt's Infitah: 1971–80', *Social Problems*, 28(4), 394–409.

—— and Tignor, R. (1982) *The Political Economy of Income Distribution in Egypt*, New York/London: Homes & Meier.

Abu-Lughod, J. (1961) 'Migrant Adjustment to City Life: The Egyptian Villages in Cairo', *Ekistics*, vol. 25, 192–202.

—— (1989) *Die Wirkungen internationaler Migration auf das Wachstum von Kairo und die Urbanisierung der ländlichen Gebiete Ägyptens*, Mimeo, Berlin: Berliner Institut für vergleichende Sozialforschung.

Adams, R. (1986) *Development and Social Change in Rural Egypt*, Syracuse: Syracuse University Press.

—— (1987) 'The Use of Workers' Remittances in Rural Development', Unpublished ms., Cairo.

—— (1988) 'Worker Remittances and Inequality in Rural Egypt', *Economic Development and Cultural Change* (Oct.)

Agwah, A. (1978) 'Import Distribution, Export Expansion and Consumption Liberalization: The Case of Egypt', *Development and Change* 9(2), 299–329.

Aly, F. (1986) 'The Socio-economic Impact of Remittances of the Egyptians Working Abroad on the Egyptian Economy from 1970–1983' (in Arabic), *al-ʿUlum al-Ijtimaʿiyya* 14(1), 71–94.

Amin, G.A. (1981) 'Some Economic and Cultural Aspects of Economic Liberalization in Egypt' *Social Problems* 28(4), 430–41.

—— and Awny, E. (1985) *International Migration of Egyptian Labour. A Review of the State of the Art*, Ottawa: International Development Research Centre.

Ammar, H. (1954) *Growing up in an Egyptian Village*, London: Routledge.

Anderson, B. (1983) *Imagined Communities: Reflections on the Origin and Spread of Nationalism*, London: Verso.

Ansari, H. (1987) *Egypt, the Stalled Society*, Cairo: The American University in Cairo Press.

Appadurai, A. (1990) 'Disjuncture and Difference in the Global Cultural Economy', *Theory, Culture and Society*, vol. 7, 295–310.

Appleyard, R. (1989) 'International Migration and Developing Countries', in Appleyard, R. (ed.), *The Impact of International Migration On Developing Countries*, Paris: OECD.

Arbeitsgruppe Bielefelder Entwicklungssoziologen (eds) (1981) *Subsistenzproduktion und Akkumulation*, Saarbrücken and Fort Lauderdale: Breitenbach.

Arnold, F. and Shah, N.M. (1986) 'Asia's Labor Pipeline: An Overview', in Arnold, F. and Shah, N.M. (eds), *Asian Labor Migration. Pipeline to the Middle East*, Boulder, Colo. and London: Westview Press.

Ayrout, H. (1963) *The Egyptian Peasant*, Boston.

Ayubi, N. (1982) 'Implementation Capabilities and Political Feasibility of the Open-Door Policy', in Kerr, M.H. and Yassin S. (eds), *Rich and Poor States in the Middle East, Egypt and the New Arab Order*, Cairo: The American University in Cairo Press.

Bach, R.L. and Schraml L.A. (1982) 'Migration, Crisis, and Theoretical Conflict', *International Migration Review* 16(2), 320–41.

Baer, G. (1962) *A History of Landownership in Modern Egypt, 1800–1950*. London: Oxford UP.

—— (1967) 'Slavery in 19th Century Egypt', *Journal of African History* 8(3), 417–41.

—— (1969) 'Slavery and its Abolition', in Baer G., *Studies in the Social History of Modern Egypt*, Chicago, Ill.: Chicago University Press.

Bassiouni, M. Sh. (1988/89) 'Egypt in Transition: Perspectives on a Rapidly Changing Society', *American Arab Affairs* 27, 70–96.

Baudrillard, J. (1972) *Pur une critique de l'oéconomie politique du signe*, Paris: Gallimard.

—— (1973) *Le miroir de la production*, Tournai: Casterman.

Beaugé G. and Büttner, F. (eds) (1991) *Les migrations dans le monde arabe*, Paris: CNRS.

Beck, L. and Keddie, N. (1978) *Women in the Muslim World*, Cambridge, Mass.: Harvard UP.

Bendiab, A. (1989) 'Labour Migration of Egyptian Women to the Gulf States', *Nord-Süd aktuell*, no. 4, 540–43.

—— (1991) 'Femmes et migration vers les pays du Golfe-Remarques sur l'état de la recherche', in Beaugé G. and Büttner, F. (eds), *Les migrations dans le monde arabe*, Paris: CNRS.

Bennholdt-Thomsen, V. (1980) 'Investition in die Armen. Zur Entwicklungspolitik der Weltbank', *Lateinamerika, Analysen und Berichte*, vol. 4, 74–96.

Bergo, L. (1985) 'Emigration, Etat et Population en Egypte. Une Hypothèse d'interprétation', *Peuples Méditerranéens* no. 31/32, 61–7.

Bernal, V. (1985) 'Household Agricultural Production and Off-Farm Work in a Blue Nile Village, Sudan', Evanston, Ill.: Northwestern University Dissertation (photocopy).

—— (1991) *Cultivating Workers, Peasants and Capitalism in a Sudanese Village*, New York: Columbia University Press.

Birks, J.S. and Sinclair, C.A. (1980) *International Migration and Development in the Arab Region*, Geneva: ILO.

—— (1989) 'International Migration for Employment. Manpower and Population Evolution in the GCC and the Libyan Arab Jamahiriya', World Employment Programme Research, Working Paper 42, Geneva, International Labour Office.

——, Serageldim, I., Sinclair, C.A. and Socknat, J.A. (1983a) 'Who is Migrating Where? An Overview of International Labour Migration in the Arab World', in Richards, A. and Martin, Ph.L., *Migration, Mechanization, and Agricultural Labor Markets in Egypt*, Boulder, Colo. and Cairo: Westview Press.

——, Sinclair, C.A. and Socknat, J.A. (1983b) 'The Demand for Egyptian Labor Abroad', in Richards, A. and Martin, Ph.L., *Migration, Mechanization, and Agricultural Labor Markets in Egypt*, Boulder, Colo.: Westview Press.

——, Seccombe, I.J. and Sinclair, C.A. (1986) 'Migrant Workers in the Arab Gulf: The Impact of Declining Oil Revenues', *International Migration Review* XX(2,4), 799–814.

Blaschke, J. (1989) 'Gewalt und Wanderungen. Der Weltmarkt für Arbeit', *Der Überblick*, no. 4, 5–10.

Bliss, F. (1984/5) 'Traditionelle Gesellschaft, Regionalentwicklung und nationaler Rahmen in Ägypten', *Sociologus* 34(1), 97–120.

Böhning, W.R. (1984) *Studies in International Labour Migration*, London: Macmillan.

Boinet, A. (1899) *Dictionaire Géographique de l'Égypte*, le Caire: Ministry of Finance.

Bourdieu, P. (1977) *Outline of a Theory of Practice*, Cambridge: CUP.

—— (1983) 'Ökonomisches Kapital, kulturelles Kapital, soziales Kapital', in Kreckel, R., *Soziale Ungleichheiten, Sonderband 2 der Sozialen Welt*, Göttingen: Schwarz.

Boustani, R. and Fargues Ph. (1990) *Atlas du Monde Arabe Géopolitique et Société*, Paris: Bordas.

Boyne, R. (1990) 'Culture and the World-System', *Theory, Culture & Society*, vol. 7, 57–62.

Brink, J.H. (1991) 'The Effect of Emigration of Husbands on the Status of their Wives: An Egyptian Case', *International Journal of Middle East Studies* 23(2), 201–11.

Brown, N.J. (1990) *Peasant Politics in Modern Egypt. The Struggle Against the State*, New Haven, Conn. and London: Yale UP.

CAPMAS (1985) *A Statement on the Population of the Arab Republic of Egypt*, Mimeo, Cairo.
—— (1987a) 'Agriculture, Cropped Area', in National Bank of Egypt, *Economic Bulletin* XL (1,2).
—— (1987b) 'Main Groups of Exports, Exports of Petroleum and Raw Materials', in National Bank of Egypt, *Economic Bulletin* XL(4).
—— (1988) *Preliminary Results of the 1986 National Survey*, Cairo: CAPMAS.
Choucri, N. (1977) 'The New Migration in the Middle East: A Problem for Whom?', *International Migration Review* 11(4), 421–43.
—— (1986) 'Asians in the Arab World: Labor Migration and Public Policy', *Middle Eastern Studies* 22(2), 252–73.
Cohen, R. (1987) *The New Helots, Migrants in the International Division of Labour*, Aldershot, England and Brookfield, Ut.: Gower.
Commander, S. (1987) *The State and Agricultural Development in Egypt since 1973*, London: Ithaca Press.
—— and Hadhoud, A.A. (1986) 'From Labour Surplus to Labour Scarcity? The Agricultural Labour Market in Egypt', *Development Policy Review*, 4(2), 161–80.
Cooper, M. (1982) *The Transformation of Egypt*, London: Croom Helm.
Crow, G. (1989) 'The Use of the Concept of "Strategy" in Recent Sociological Literature', *Sociology* 23(1): 1–24.
Cuno, K. (1980) 'The Origins of Private Ownership of Land in Egypt: A Reappraisal', *International Journal of Middle East Studies*, 12(3), 245–75.
Dessouki, A.H. (1981) 'Policy Making in Egypt: A Case Study of the Open Door Economic Policy', *Social Problems* 28(4), 410–16.
—— (1982) 'The Shift in Egypt's Migration Policy: 1952–1978', *Middle Eastern Studies*, vol. 18, 53–68.
Dethier, J-J. (1989) *Trade, Exchange Rate, and Agricultural Pricing Policies in Egypt*. Vol. 1: *The Country Study*, World Bank Comparative Studies: The Political Economy of Agricultural Pricing Policy, Washington, DC.
Elwert, G. (1989) 'Ethnologische Artefakte und die theoretische Aufgabe der empirischen Sozialwissenschaften', *Saeculum* 40(2), 149–60.
—— and Wong, D. (1981) 'Thesen zum Verhältnis von Subsistenzproduktion und Warenproduktion in der Dritten Welt', in Arbeitsgruppe Bielefelder Entwicklungssoziologen (eds), *Subsistenzproduktion und Akkumulation*, Saarbrücken and Fort Lauderdale: Breitenbach.
Engelbrektsson, U.B. (1978) *The Force of Tradition: Turkish Migrants at Home and Abroad*, Goeteborg: Acta Universitatis Gothoburgensis.
Evans, G. (1986) *From Moral Economy to Remembered Village. The Sociology of James C. Scott*, Sociology Department, LaTrobe University, Tasmania, Australia, Working Paper no. 40.
Evers, H.D. (1987) 'Subsistenzproduktion, Markt und Staat. Der sogenannte Bielefelder Verflechtungsansatz', *Geographische Rundschau* 39(3), 136–40.
—— and Schiel, T. (1981) 'Expropriation der unmittelbaren Produzenten oder Ausdehnung der Subsistenzwirtschaft – Thesen zur bäuerlichen und städtischen Subsistenzproduktion', in Arbeitsgruppe Bielefelder Entwicklungssoziologen (eds), *Subsistenzproduktion und Akkumulation*, Saarbrücken and Fort Lauderdale: Breitenbach.

——, Claus, W. and Wong, D. (1984) 'Subsistence Reproduction: A Framework for Analysis', in Smith, J., Wallerstein, I. and Evers, H.D. (eds), *Households and the World-Economy*, Beverly Hills: Sage.

Fanon, F. (1967) *The Wretched of the Earth*, Harmondsworth: Penguin.

Fargues, Ph. (1991) 'La migration obéit-elle à la conjoncture pétrolière dans le Golfe? L'exemple du Kuweit', in Beaugé G. and Büttner, F. (eds), *Les migrations dans le monde arabe*, Paris: CNRS.

Featherstone, M. (1987) 'Consumer Culture, Symbolic Power and Universalism', in Stauth, G. and Zubaida, S. (eds), *Mass Culture, Popular Culture, and Social Life in the Middle East*, Frankfurt and Boulder, Colo.: Campus.

—— (1990) 'Global Culture: An Introduction, *Theory, Culture and Society*, vol. 7, 1–14.

—— (1991) *Consumer Culture and Postmodernism*, London: Sage.

Fergany, N. (1985) 'Migrations inter-arabes et dévelopment', *Revue Tiers Monde* XXVI(103), 583–96.

—— (1992) 'Arab Labor Migration And The Gulf Crisis', Unpublished paper presented at the AUC Conference on Socio-economic and Political Impact of the Gulf Crisis, Cairo.

Foster-Carter, A. (1978a) 'Can We Articulate "Articulation"?' in Clammer J. (ed.), *The New Economic Anthropology*, London: Macmillan.

—— (1978b) 'The Modes of Production Controversy', *New Left Review*, no. 107, 47–77.

Friedman, J. (1990) 'Being in the World: Globalization and Localization', *Theory, Culture & Society*, vol. 7, 311–28.

Friedman, K. (1984) 'Households as Income-Pooling Units', in Smith, J., Wallerstein, I. and Evers, H.D. (eds), *Households and the World-Economy*, Beverly Hills: Sage.

Friedmann, H. (1980) 'Household Production and the National Economy: Concepts for the Analysis of Agrarian Formations', *Journal of Peasant Studies* 7(3), 158–84.

Friedmann, J. (1986) 'The World City Hypothesis', *Development and Change*, vol. 17, 69–83.

Fröbel, F., Heinrichs, J. and Kreye, O. (1977) *Die neue internationale Arbeitsteilung*, Hamburg: Reinbek.

Gadalla, S.M. (1978) *Is There Hope? Fertility and Family Planning in a Rural Egyptian Community*, Chapel Hill: Carolina Population Center.

Geertz, C. (1983) *Local Knowledge: Further Essays on Interpretive Anthropology*, New York: Basic Books.

Giddens, A. (1977) *Studies in Social and Political Theory*, London: Hutchison; New York: Basic Books.

—— (1991) 'Structuration Theory: Past, Present and Future', in Bryant C.G.A. and Jary, D. (eds), *Giddens' Theory of Structure. A Critical Appreciation*, London: Routledge.

Glavanis, K. and Glavanis, P. (1983) 'The Sociology of Agrarian Relations in the Middle East: The Persistence of Household Production', *Current Sociology* 31(2), 1–109.

—— (eds) (1990) *The Rural Middle East. Peasant Lives and Modes of Production*, London and New Jersey: Zed Press.

Gran, J. (1977) 'Impact of the World Market on Egyptian Women', *MERIP Report*, no. 95, 3–9.

Hammam, M. (1981) 'Labour Migration and the Sexual Division of Labour', *Middle East Report*, no. 124, 5–11.

Hansen, B. and Radwan, S. (1982) *Employment Opportunities and Equity in a Changing Economy: Egypt in the 1980s, a Labour Market Approach*, Geneva: ILO.

Hareven, T. (1974) 'The Family as Process: The Historical Study of the Family Cycle', *Journal of Social History* VII(3), 322–9.

—— (1975) 'Review Essay: Household and Family in Past Time', *Theory and History* XIV(2), 242–51.

Hélie-Lucas, M-A. (1989) 'Review: Bouthaina Shaaban: Both Right And Left Handed. Arab Women Talk About Their Lives', *Beiträge zur feministischen Theorie und Praxis*, no. 25/26, 237–41.

Hobsbawm, E. and Ranger, T. (eds) (1983) *The Invention of Tradition*, Cambridge: CUP.

Hooglund, E. (1991) 'The Other Face of War', *Middle East Report*, vol. 21, no. 171(4), 3–7, 10–12.

Hopkins, N. (1988) *Agrarian Transformation in Egypt*, Cairo: The American University in Cairo Press.

Hopkins, T.K. and Wallerstein, I. (1979) 'Grundzüge der Entwicklung des modernen Weltsystems. Entwurf eines Forschungsvorhabens', in Senghaas, D. (ed.), *Kapitalistische Weltökonomie, Kontroversen über ihren Ursprung und ihre Entwicklungsdynamik*, Frankfurt am Main: Suhrkamp.

Ibrahim, S.E. (1982a) 'Oil, Migration and the New Arab Social Order', in Kerr, M.H. and Yassin, S. (eds), *Rich and Poor States in the Middle East, Egypt and the New Arab Order*, Cairo: The American University in Cairo Press.

—— (1982b) *The New Arab Social Order: A Study of the Oil Wealth*, Boulder, Colo.: Westview Press.

—— (1987) 'A Sociological Profile', in Saqqaf, Y., *The Middle Eastern City. Ancient Traditions Confront a Modern World*, New York: Paragon House Publishers.

ILO (1982) *Employment Opportunities and Equity in Egypt*, Geneva: ILO.

Kandiyoti, D. (1977) 'Sex Roles and Social Change: A Comparative Appraisal of Turkey's women', *Sign* 3(1), 57–73.

—— (1990) 'Women and Household Production: The Impact of Rural Transformation in Turkey', in Glavanis, K. and Glavanis, P. (eds), *The Rural Middle East. Peasant Lives and Modes of Production*, London and New Jersey: Zed Press.

Kerr, M.H. and Yassin, S. (eds) (1982) *Rich and Poor States in the Middle East. Egypt and the New Arab Order*, Cairo: The American University in Cairo Press.

Khafagy, F. (1983) 'Socio-Economic Impact of Emigration from a Giza Village', in Richards, A. and Martin, Ph.L., *Migration, Mechanization, and Agricultural Labor Markets in Egypt*, Boulder, Colo. and Cairo: Westview Press.

—— (1984) 'Women and Labor Migration: One Village in Egypt', *Middle East Report*, no. 124, 14–21.

Khan, K.M. (1989) 'Internationale Migration und Entwicklungsländer', *Nord-Süd aktuell*, no. 4, 497–505.

—— (1991) 'Les migration internationales vers le Moyen-Orient et leur impact sur l'économie pakistanaise', in Beaugé G. and Büttner, F. (eds), *Les migrations dans le monde arabe*, Paris: CNRS.

Khattab, H.A.S. and El Daeif, S.G. (1982) *Impact of Male Labor Migration on the Structure of the Family and the Roles of Women*, Cairo: The Population Council of West Asia and North Africa.

Kurayyam, Ḳ. (1986) 'al-athar al-iqtisadiyya li-hijrat al-ʿumala ʿalar-rif al-misri', *L'Égypte contemporaine*, no. 404: 181–98.

Labib, A. (1991) 'L'immigration asiatique dans les pays du Golfe', in Beaugé G. and Büttner, F. (eds), *Les migrations dans le monde arabe*, Paris: CNRS.

Lambert, A. (1943) 'Les Salariés Dans l'Entreprise Agricole Egyptienne', *L'Égypte contemporaine*, no. 33, 223–35.

Lenin, V.I. (1967) *The Development of Capitalism in Russia* (Published according to the text of the second edition, 1908), Moscow: Foreign Languages Publishing House.

Lerner, D. (1958) *The Passing of Traditional Society*, Glencoe: Free Press.

Lozach, J. and Hug. G. (1930) *L'habitat rural en Égypte*, Cairo: Société de géographie d'Égypte.

Luxemburg, R. (1969) *Die Akkumulation des Kapitals*, Frankfurt: Neue Kritik.

Macleod, A.E. (1990) *Accommodating Protest: Working Women, the New Veiling, and Change in Cairo*, New York: Columbia UP.

Maher, V. (1974) *Women and Property in Morocco: Their Changing Relation to the Process of Social Stratification in the Middle Atlas*, London: CUP.

Mannheim, K. (1957) *Systematic Sociology. An Introduction to the Study of Sociology*, London: Routledge.

Meillassoux, C. (1972) 'From Production to Reproduction', *Economy and Society*, no. 1: 92–105.

—— (1976) *Die wilden Früchte der Frau*, Frankfurt am Main: Suhrkamp.

Messiri, S. (1983) 'Tarahil Laborers in Egypt', in Richards, A. and Martin, Ph.L., *Migration, Mechanization, and Agricultural Labor Markets in Egypt*, Boulder, Colo. and Cairo: Westview Press.

The Middle East and North Africa 1990 (1989), London: Europa Publications Limited.

Miller, J. (1985) 'End of Oil Boom in Gulf Sends Arab Migrant Workers Home to Uncertain Future', *International Herald Tribune*, Monday Oct. 7, p. 2.

Mitchell, T. (1988) *Colonizing Egypt*, Cambridge: Cambridge UP.

—— (1991) 'America's Egypt: Discourse of the Development Industry', *Middle East Report*, vol. 21, no. 169(2), 18–34.

Myntti, C. (1984) 'Yemeni Workers Abroad: The Impact on Women', *Middle East Report* (June), pp. 11–16.

Nada, A.H. (1991) 'Impact of Temporary International Migration on Rural Egypt', *Cairo Papers in Social Science, Social Research Center Research Series* 14(3), 1–76.

Nahas, J.F. (1901) *Situation économique et sociale du fellah égyptien*, Paris: A. Rousseau.

OECD (1987) *The Future of Migration*, Paris: OECD.

Otto-Walter, R. (1981) 'Unterentwicklung und Subsistenzproduktion – Forschungsansatz der Arbeitsgruppe Bielefelder Entwicklungssoziologen', in Arbeitsgruppe Bielefelder Entwicklungssoziologen (eds), *Subsistenzproduktion und Akkumulation*, Saarbrücken and Fort Lauderdale: Breitenbach.

Owen, R. (1981a) 'The Development of Agricultural Production in Nineteenth Century Egypt: Capitalism of What Type?', in Udovitch, A.L. (ed.), *The Islamic Middle East: Studies in Economic and Social History*, Princeton, N.J.: The Darwin Press.

—— (1981b) *The Middle East in The World Economy 1800–1914*, London: Methuen.

—— (1985) *Migrant Labour in the Gulf*, London: Minority Rights Group.

—— (1986) 'Large Landowners, Agricultural Progress and the State in Egypt, 1800–1970: An Overview', in Richards, A. (ed.), *Food, States, and Peasants, Analyses of the Agrarian Question in the Middle East*, Boulder, Colo. and London: Westview Press.

Pahl, R.E. (1985) 'The Restructuring of Capital, the Local Political Economy and Household Work Strategies', in Gregory, D. and Urry, J. (eds), *Social Relations and Spatial Structures*, London: Macmillan.

Park, S.J. (1991) 'La stratégie d'exportation de main-d'oeuvre des multinationales asiatiques vers la Golfe: l'optimum S/Q.', in Beaugé G. and Büttner, F. (eds), *Les migrations dans le monde arabe*, Paris: CNRS.

Paye, J-C. (1987) 'Introduction to the Conference', in OECD, *The Future of Migration*, Paris: OECD.

Popkin, S. (1979) *The Rational Peasant: The Political Economy of Rural Society in Vietnam*, Berkeley, Calif.: University of California Press.

Portes, A. and Walton, J. (1981) *Labor, Class, and the International System*, New York: Academic Press.

Potts, L. (1988) *Weltmarkt für Arbeit. Von der Kolonisation Amerikas bis zu den Migrationen der Gegenwart*, Hamburg: Junius.

—— (1990) *The World Labour Market. A History of Migration*, London: ZED Books.

Radwan, S. (1977) *Agrarian Reform and Rural Poverty, 1952–1975*, Geneva: International Labour Office.

Ramzi, M. (1954) *al-qamus al-jugrafi li'l-bilad al-misriyya min 'ahd qudama' al-misriyyin ila sanat 1945*, vols I–IV, Cairo: Dar al-kutub al misriyya.

Reichert, Ch. (1990a) 'Labour Migration and Rural Development in Egypt. A Study of Return Migration in Six Villages', Paper presented to the 14th European Congress of Rural Sociology, Giessen.

—— (1990b) 'Auswirkungen der Arbeitsmigration im ländlichen Ägypten – Eine Rückkehreruntersuchung in sechs Dörfern' (1987/1988) in (ed)

Berliner Institut für Vergleichende Sozialforschung e. V., *Jahrbuch für Vergleichende Sozialforschung*: 57–100.
Richards, A. (1978) 'Land and Labor on Egyptian Cotton Farms, 1882–1940', *Agricultural History* 25(3), 503–18.
—— (1979) 'The Political Economy of Gutswirtschaft: A Comparative Analysis of East Elbian Germany, Egypt and Chile', *Comparative Studies in Society and History* vol. 21, 483–518.
—— (1982) *Egypt's Agricultural Development, 1800–1980. Technical and Social Change*, Boulder, Colo.: Westview Press.
—— (ed.) (1986) *Food, States, and Peasants, Analyses of the Agrarian Question in the Middle East*, Boulder, Colo. and London: Westview Press.
—— (1991) 'Agricultural Employment, Wages and Government Policy in Egypt During and After the Oil Boom', in Handoussa, H. and Potter, G. (eds), *Employment and Structural Adjustment, Egypt in the 1990s*, Cairo: The American University in Cairo Press.
—— and Martin, Ph.L. (1983) *Migration, Mechanization, and Agricultural Labor Markets in Egypt*, Boulder, Colo. and Cairo: Westview Press.
—— and Waterbury, J. (1990) *A Political Economy of the Middle East, State, Class, and Economic Development*, Cairo: The American University in Cairo Press.
Rivlin, H. (1961) *The Agricultural Policy of Muhammad Ali in Egypt*, Cambridge, Mass.: Harvard UP.
Rivlin, P. (1985) *The Dynamics of Economic Policy Making in Egypt*, New York: Praeger.
Roy, D.A. (1991) 'Egyptian Emigrant Labor: Domestic Consequences', *Middle Eastern Studies* 27(4), 551–82.
Rugh, A.B. (1985) *Family in Contemporary Egypt*, Cairo: The American University in Cairo Press.
—— (1989) *Reveal and Conceal. Dress in Contemporary Egypt*, Cairo: The American University in Cairo Press.
Saab, G. (1967) *The Egyptian Agrarian Reform 1952–1962*, London: Oxford University Press.
Sabagh, G. (1982) 'Migration and Social Mobility in Egypt', in Kerr, M.H. and Yassin, S. (eds), *Rich and Poor States in the Middle East, Egypt and the New Arab Order*, Cairo: The American University in Cairo Press.
Sadat, A. (1978) *In Search of Identity. An Autobiography*, New York: Harper & Row.
Saffa, S. (1949) *Exploitation économique et agricole d'une domaine rural égyptien*, Le Caire: Thèse.
Said, M.S. (1989) The Political Economy of Migration in Egypt 1974–1989, Cairo: Center for Political and Strategic Studies: al-Ahram.
Sassen, S. (1988) *The Mobility of Labour and Capital. A Study in International Investment and Labor Flow*, Cambridge: CUP.
Sassen-Koob, S. (1983) 'Labour Migrations and the New International Division of Labor', in Nash, J. and Fernandez-Kelly, M.P. (eds), *Women, Men and the International Division of Labor*, Albany, N.Y.: State University of New York Press.

Schiel, T. and Stauth, G. (1981) 'Subsistenzproduktion und Unterentwicklung', *Peripherie*, no. 5/6, 122–43.

Scott, J.C. (1976) *The Moral Economy of the Peasant*, New Haven, Conn.: Yale UP.

—— (1985) *Weapons of the Weak. Everyday Forms of Peasant Resistance*, New Haven, Conn.: Yale UP.

Seccombe, I.J. (1988) 'International Migration in the Middle East: Historical Trends, Contemporary Patterns and Consequences', in Appleyard, R.T. (ed.), *International Migration Today, Vol. 1: Trends and Prospects*, University of Western Australia, Centre for Migration and Development Studies: UNESCO.

—— and Findlay, A.M. (1989) 'The Consequences of Temporary Emigration and Remittance Expenditure from Rural and Urban Settlements: Jordan', in Appleyard R. (ed.), *The Impact of International Migration On Developing Countries*, Paris: OECD.

—— and Lawless, R.I. (1989) 'State Intervention and the International Labour Market: A Review of Labour Emigration Policies in the Arab World', in Appleyard R. (ed.), *The Impact of International Migration on Developing Countries*, Paris: OECD.

Sell, R. (1987) 'Gone for Good', *Cairo Papers in Social Science*, vol. 10, monograph 2, 1–97.

Senghaas, D. (ed.) (1972) *Imperialismus und strukturelle Gewalt*, Frankfurt am Main: Suhrkamp.

—— (ed.) (1979) *Kapitalistische Weltökonomie. Kontroversen über ihren Ursprung und ihre Entwicklungsdynamik*, Frankfurt am Main: Suhrkamp.

Shah, N.M. and Al-Qudsi, S.S. (1989) 'The Changing Characteristics of Migrant Workers in Kuwait', *International Journal of Middle East Studies* 21(1), 31–55.

Smith, J., Wallerstein, I. and Evers, H.D. (eds) (1984) *Households and the World-Economy*, Beverly Hills: Sage.

Statistisches Bundesamt (1983) *Länderbericht Ägypten* Wiesbaden: Statistisches Bundesamt.

Stauth, G. (1981) 'Zur Auflösung und Peripherisierung integrierter Kleinbauernwirtschaft – Das Beispiel der Fellachen im Nildelta', in Arbeitsgruppe Bielefelder Entwicklungssoziologen (eds), *Subsistenzproduktion und Akkumulation*, Saarbrücken and Fort Lauderdale: Breitenbach.

—— (1983) *Die Fellachen im Nildelta*, Wiesbaden: Steiner.

—— (1984) 'Households, Modes of Living, and Production Systems', in Smith, J., Wallerstein, I. and Evers, H.D. (eds), *Households and the World-Economy*, Beverley Hills: Sage.

—— (1990a) 'Die Modernität des Dorfes', in Stauth, G., Albrecht, R., Reichert, Ch. and Weyland, P., 'Fellachen zwischen Überlebenssicherung und Modernität als Lebensstil. Strukturelle Unterentwicklung, Migration, Geschlechterverhältnis und Islamisierung: Die Konflikte der ländlichen Gesellschaft Ägyptens, 1986–1988', Unpubl. ms., Bielefeld.

—— (1990b) 'Mittelklasse und Islam', in Stauth, G., Albrecht, R., Reichert, Ch. and Weyland, P., 'Fellachen zwischen Überlebenssicherung und Modernität als Lebensstil. Strukturelle Unterentwicklung, Migration, Geschlechterverhältnis und Islamisierung: Die Konflikte de ländlichen Gesellschaft Ägyptens, 1986–1988', Unpubl. ms., Bielefeld.

—— (1990c) 'Theoretische Vorbemerkungen zu einer Soziologie des Konflikts zwischen kultureller Dynamisierung und lokaler Verfügung ländlicher Ressourcen, in Stauth G., Albrecht, R., Reichert, Ch. and Weyland, P., 'Fellachen zwischen Überlebenssicherung und Modernität als Lebensstil. Stukturelle Unterentwicklung, Migration, Geschlechterverhältnis und Islamisierung: Die Konflikte der ländlichen Gesellschaft, Ägyptens, 1986–1988, Unpublished ms., Bielefeld.

—— (1990d) 'Capitalist Farming and Small Peasant Housholds in Egypt' in Glavanis, K. and Glavanis, P., *The Rural Middle East. Peasant Lives and Modes of Production*, London and New Jersey: Zed Press.

—— and Zubaida, S. (eds) (1987) *Mass Culture, Popular Culture, and Social Life in the Middle East*, Frankfurt and Boulder, Colo.: Campus.

—— and Turner, B.S. (1988) 'Nostalgia, Postmodernism and the Critique of Mass Culture', *Theory, Culture and Society*, vol. 5, 509–26.

—— and Weyland, P. (1990) 'Transformationen ländlicher Gesellschaft und sozialer Konflikt: Das Beispiel Ägypten', in Stauth, G., Albrecht, R., Reichert, Ch. and Weyland, P., 'Fellachen zwischen Überlebenssicherung und Modernität als Lebensstil. Strukturelle Unterentwicklung, Migration, Geschlechterverhältnis und Islamisierung: Die Konflikte der ländlichen Gesellschaft Ägyptens, 1986–1988', Unpubl. ms., Bielefeld.

Sukkary-Stolba, S. (1985) 'Changing Roles of Women in Egypt's Newly Reclaimed Lands', *Anthropological Quarterly* 58(4), 182–9.

Sullivan, E.L. and Korayem, K. (1981) 'Women and Work in the Arab World', *Cairo Papers in Social Sciences*, vol. 4, monograph 4.

Taylor, E. (1984) 'Egyptian Migration and Peasant Wives', *MERIP Reports*, no. 124, 3–10.

Thurley, K. and Wood, S. (1983) 'Introduction', in Thurley, K. and Wood, S. (eds), *Industrial Relations and Management Strategy*, Cambridge: CUP.

Toth, J. (1991) 'Pride, Purdah, or Paychecks: What Maintains the Gender Division of Labor in Rural Egypt?', *International Journal of Middle East Studies* 23(2), 213–36.

Tucker, J. (1979) 'Decline of the Family Economy in Mid-nineteenth-century Egypt', *Arab Studies Quarterly* 1(3), 245–71.

—— (1986) *Women in Nineteenth-century Egypt*, Cairo: The American University in Cairo Press (first publ. by Cambridge University Press in 1985).

Udovitch, A.L. (ed.) (1981) *The Islamic Middle East: Studies in Economic and Social History*, Princeton, N.J.: The Darwin Press.

Wallace, D.M. (1883) *Egypt and the Egyptian Question*, London: Macmillan and Co.

Wallerstein, I. (1974) *The Modern World System*. Vol. I: *Capitalist Agriculture and the Origins of the European World Economy in the 16th Century*, New York: Cambridge University Press.

—— (1979) 'Aufstieg und künftiger Niedergang des kapitalistischen Weltsystems. Zur Grundlegung vergleichender Analyse', in Senghaas, D. (ed.), *Kapitalistische Weltökonomie, Kontroversen über ihren Ursprung und ihre Entwicklungsdynamik*, Frankfurt am Main: Suhrkamp.

—— (1984) 'Household Structures and Labor-Force Formation in the Capitalist World-Economy', in Smith, J., Wallerstein, I. and Evers, H.D. (eds), *Households and the World-Economy*, Beverly Hills: Sage.

—— (1986) *Das moderne Weltsystem – Die Anfänge kapitalistischer Landwirtschaft und die europäische Weltökonomie im 16, Jahrhundert*, Frankfurt am Main: Syndikat.

Walz, T. (1985) 'Black Slavery in Egypt during the Nineteenth Century as Reflected in the Mahkama Archives of Cairo', in Willis, J.R. (ed.), *Slaves and Slavery in Muslim Africa* (vol. II), London: Cass.

Warriner, D. (1962) *Land Reform and Development in the Middle East. A Study of Egypt, Syria, and Iraq*, London: Oxford UP.

Waterbury, J. (1985) '"The Soft State" and the Open Door: Egypt's Experience with Liberalization 1974–1984', *Comparative Politics* (Oct), 65–83.

Watzlawick, P. (1985) 'Selbsterfüllende Prophezeiungen', in Watzlawick, P. (ed.), *Die erfundene Wirklichkeit*, München: Piper.

Werlhof, C. von (1985) *Wenn die Bauern wiederkommen*, Bremen: Ed. Con.

—— and Neuhoff, H.P. (1979) 'Zur Logik der Kombination verschiedener Produktionsverhältnisse – Beispiele aus dem venezolanischen Agrarsektor', *Lateinamerika: Analysen und Berichte*, vol. 3, 86–117.

Weyland, P. (1987) 'Remigration und Geschlechterverhältnis auf dem ägyptischen Dorf. Forschungsperspektiven und Hypothesen', Working Paper no. 95, Bielefeld.

Wirz, A. (1984) *Sklaverei und kapitalistisches Weltsystem*, Frankfurt am Main: Suhrkamp.

Wong, D. (1987) *Peasants in the Making Malaysia's Green Revolution*, Singapore: ISEAS.

Wood, C.H. (1981) 'Structural Changes and Household Strategies: A Conceptual Framework for the Study of Rural Migration', *Human Organization*, no. 404, 338–44.

—— (1982) 'Equilibrium and Historical Structural Perspectives on Migration', *International Migration Review* 16(2), 298–319.

World Labour Report 1 (1984) *Employment, Incomes, Social Protection, New Information Technology*, Geneva: ILO.

Zimmermann, S. (1982) *The Women of Kafr al-Bahr, A Research into the Working Conditions of Women in an Egyptian Village*, Cairo and Leiden: Research Centre, Women and Development, State University of Leiden, Institute for Social Studies.

INDEX

Note: Page numbers in *italic* type refer to tables. Numbers in brackets refer to notes.